Receiving Antennas
for the Radio Amateur

Eric P. Nichols, KL7AJ

Production
Michelle Bloom, WB1ENT
Sue Fagan, KB1OKW — Cover Art
David F. Pingree, N1NAS

Our Cover: Two views of the active Beverage receiving antenna installed at OT5A, the contest station of Lierse Radioamateurs near Antwerp, Belgium. Photos courtesy of Lierse Radioamateurs.

Copyright © 2018 by

The American Radio Relay League, Inc.

Copyright secured under the Pan-American Convention

All rights reserved. No part of this work may be reproduced in any form except by written permission of the publisher. All rights of translation are reserved.

Printed in USA

Quedan reservados todos los derechos

ISBN: 978-1-62595-078-9

First Edition
Second Printing

We strive to produce books without errors. Sometimes mistakes do occur, however. If you think you have found an error, please let us know by sending e-mail to **pubsfdbk@arrl.org**.

*O what a tangled web we weave,
When first we practice to receive…*

— Otto Watt, September 5, 1925

Contents

Preface

Introduction: The Limits of Reciprocity

About the ARRL

1 Receiving Antennas are Different!

2 Your Friend, the Decibel

3 The Preamplifier Problem

4 The Amazing Disappearing Antenna

5 The Receiving Antenna as a Signal Generator

6 The Quest for Infinite Power Gain

7 The Role of the Resistor in the Receiving Antenna

8 The Small Loop Antenna

9 Achieving the Perfect Null

10 You *Can* Make Accurate Field Strength Measurements

11 The Aperiodic Loop

12 The Q Factor

13 The Beverage: In a Class of Its Own

14 Dealing with Non-Reciprocal Propagation

15 The Evolution of the eXOgon Antenna

16	The Lowdown on LF
17	The Random Wire
18	Arrays and Beamforming Networks
19	Powering Your Active Antenna
20	Diversity Methods
21	NVIS Receiving Antennas
22	Receiving Antenna Projects and Accessories
23	Materials and Construction Techniques
24	Our Two New Bands
	Appendix A — The Q of Everything
	Appendix B — Get a Load of This: Taking the Mystery Out of Loaded and Unloaded Q
	Appendix C — Broadband Loop Antenna System
	Appendix D — Low Noise Amplifiers
	Appendix E — Online Resources
	Index

Preface

Total Copper Content

This book has been simmering on the back burner for a long time. Or perhaps, *languishing* is a more apt term. After decades of studying and applying the principles of the best books on antenna theory ever written, I struggled for a while with the question of how I could possibly contribute anything new to the field. I knew it would be a daunting task to come up with any truly new information…or even a new *angle*. The task loomed even more dauntingly when I realized that a radio antenna is really nothing more than a piece of wire.

Chemistry and biology and neurology, and even computer science, are very complicated. There are lots of nooks and crannies in a molecule in which to hide all kinds of nuances. There are trillions of synaptic connections in a human brain. Even a low-end microprocessor has millions of connections that can be wired in countless different ways. You could spend a lifetime just getting started in any of these fields.

But an antenna is a piece of wire. There's nowhere for a nuance to hide. Not a nook nor a cranny to be found. It would seem infantile to spend a lifetime studying a piece of wire.

And yet, many people a lot smarter than I have spent their lifetimes doing precisely that. Only most unpoetic and prosaic of hams hasn't sensed something special afoot upon connecting a scraggly bit of wire to the antenna terminal of a shortwave receiver. What is it about this humble piece of wire that allows it to intercept the still small voice of a wobbling electron hundreds or thousands of miles away?

Some time back some of us "properly seasoned" hams were debating about what single factor set the performance of one Amateur Radio station above another. One of the sages present said it could be boiled down to a simple formula — Total Copper Content, (TCC). The inescapable conclusion was the more copper you had in the air, the better your station worked. I've used the term countless times since that meeting of the minds. Although most hams intuitively know that the antenna is the most important part of the ham station, I believe nothing expresses it as concisely as Total Copper Content.

Now, we understand that not all antennas are made out of copper, and that haphazardly flinging masses of metal (of any kind) into the ether is no guarantee of success, but it is at least a starting point. The next step is to understand how best to arrange that Total Copper Content for maximum radio effectiveness. And that is largely what this book is about.

In this volume we will review some antenna theory from the vantage point of a receiving antenna. We will demonstrate how the receiving antenna, paradoxically enough, is actually a signal generator. The maximum power transfer theorem applies just as thoroughly as in the transmitting case. We will demonstrate why the absolute maximum possible efficiency of a receiving antenna is only 50%, and why that's not a problem.

As we suggest in the Introduction, one of the best-known cardinal doctrines of antenna theory is the concept of reciprocity. The gain, directivity, radiation pattern, and electrical impedance of an antenna are the same, whether it's used as a transmitting antenna or receiving antenna. While on the most fundamental level, reciprocity is certainly true; in many cases it is over-applied. It can be easy to overlook the fact that transmitting and receiving antennas perform very different roles in the real world, and they are subject to very different requirements and priorities. Even in such cases where antenna reciprocity is fully applicable, we will often find that factors well outside the control of the antenna are absolutely not reciprocal. We dedicate three entire chapters in this book to the non-reciprocal and bi-refringent behaviors of the ionosphere, and ways to not only mitigate the effects of these, but to actually take advantage of them! Since the reciprocal properties of antennas have been covered thoroughly in many other excellent works, we will concentrate in this book on the theory and methods that make receiving antennas different from transmitting antennas. We will also present evidence to confirm that nearly every Amateur Radio station can benefit from separate transmitting and receiving antennas in most cases.

We hereby launch into the introductory chapters with this well-worn, but undeniable ham radio truth:

"If you can't hear 'em, you can't work 'em."

Eric Nichols, KL7AJ
North Pole, Alaska
February 2018

My QSL card, custom designed by Jim Massara, N2EST, of Hamtoons (hamtoons.net). Jim illustrated my very first *QST* article in December 1983, "Try This Speech Decompressor," as well as a number of subsequent articles. I commissioned him to do this re-creation of the classic *QST* "Jeeves" character, which, as any old timer will tell you, is absolutely impeccable. I've always considered Jeeves my true Elmer.

Introduction: The Misinterpretation and Misapplication of Reciprocity

There was a time in the not-too-distant past when the typical radio amateur's first piece of equipment was a shortwave radio receiver. Long before the prospective ham would begin studying for an actual ham "ticket" in earnest, he or she had a pretty good "feel" for long distance, shortwave radio communications. Typically, one could take it for granted that a piece of wire haphazardly tacked to a door frame or flung over a convenient tree branch would prove more than adequate to at least give shortwave listening "SWLing" a spin. Such hams in the making were more likely to be experimentally than theoretically informed, but nevertheless accomplished great things both before and after receiving their licenses. The concept of Total Copper Content (TCC) as the primary predictor of radio communications effectiveness was validated by the countless early radio texts exhorting hams to get as much wire as high into the air as possible…and by the fact that following such advice actually *worked*.

Radio amateurs (and professionals) managed to perform wonders with only the most rudimentary understanding of radio and antenna theory. Perusing some of the earliest archives of Amateur Radio antenna theory with today's understanding, one is simultaneously struck by the astounding ignorance of some of the offerings, appearing simultaneously with some of the most up-to-date, insightful technical writing ever produced. Undoubtedly, a great deal of the charm and romance of radio is precisely because hams could do so much while knowing so little. It could be argued that the exact opposite is the case now, as it seems the "paralysis of analysis" prevents many new hams from even attempting long distance communications until they have amassed a collection of technology only dreamed of by Marconi in his prime.

It is certainly safe to say that we do have more genuine, verifiable knowledge about antennas than we did at the beginning. Electromagnetic theory is now fairly well "canonized," with a few, perhaps questionable, exceptions. Antenna design is predictable to a very large degree; it had already achieved this status before the advent of any of the antenna modeling programs we now take for granted. The *ARRL Antenna Book* has been in print for many decades now, with extremely consistent theory having been presented ever since the first edition. Any new surprises in the field of antenna theory would be…well…*surprising*.

One of the *unsurprising* truths about antenna theory (and practice) has been the concept of *reciprocity*. At a very fundamental level there is no difference whatsoever between a transmitting antenna and a receiving antenna. This is a profoundly *simplifying* truth, not only when it comes to understanding antennas, but also in designing and constructing them. This reciprocity goes right down to the subatomic level, where the interaction between fields and accelerating charged particles is fully reversible in every measureable regard. On the more visible level, we can make any kind of measurement on a receiving antenna and be confident it will perform similarly as a transmitting antenna…and vice versa. Reciprocity is so universal a concept that has been a "given" in all modern antenna literature. The fact that most radio amateurs, most of the time, use the same antenna for transmitting and receiving is testimony to the power of the reciprocity theorem. It certainly has simplified…or at least *shortened*…a lot of ham radio antenna literature. It is a doctrine so powerful in ham radio circles that it has made it easy to overlook another simple but equally profound truth.

Transmitting and receiving antennas have different jobs to do.

The well-known principle of antenna reciprocity is often over-applied by radio amateurs. While it is true that the fundamental characteristics of antennas apply to both transmission and reception, the requirements and priorities of receiving antennas can be vastly different from those of transmitting antennas. The focus of this volume will be entirely on receiving antennas, both active and passive, and their associated circuits. There are relatively few cases where a radio amateur cannot benefit from a separate, well-designed receiving antenna or antenna system.

The venerable *ARRL Antenna Book* has, for decades, comprehensively covered the topic of Amateur Radio antennas from both a theoretical and practical basis. As tempting as it might be to join a growing contingent of somewhat vocal radio amateurs, (and some professionals), who claim to have some entirely new understanding of electromagnetic theory, this is not going to happen in this volume. You will find no "revolutionary" antenna theory or devices here. We are absolutely bound by the laws of classical electromagnetics and well-understood and accepted principles. All the concepts and hardware presented will be easily cross-checked and corroborated with the *ARRL Antenna Book* and other similar respectable works. What *is* new in this volume is its special emphasis on receiving antenna techniques, a topic which has been somewhat underrepresented in most Amateur Radio literature in recent years. You will also find some unique applications of modern antenna modeling methods, seldom applied to receiving antennas.

The *active* antenna holds a prominent position in this book as well, for two reasons in particular. First, some very recent developments in radio frequency semiconductors, especially low-noise radio frequency operational amplifiers, have made a number of previously difficult-to-implement active antenna designs a very simple task. Secondly, there has been a tremendous interest in recent years in the lowest frequency amateur bands. We are on the verge of obtaining two new amateur bands, at 630 and 2200 meters, respectively. As a prelude to this, numerous highly successful experiments by the ARRL 600 Meter Experimental Group band have been conducted.

Because of the current, and likely future, power restrictions that these new bands will operate under, heavy emphasis on the receiving end of these radio paths is essential. At the same time, increasingly more hams are laboring under challenging real estate and other restrictions, all of which present manifold challenges on the lower frequencies. The effective *active* antenna is one ingredient in the solution to these challenges. The timing of this volume coincides nicely with the opening of these exciting new amateur bands, allowing hams to "hit the ground running" now that these frequencies are available. We trust that this book will complement your antenna library, and serve to increase your Amateur Radio effectiveness.

About ARRL

The seed for Amateur Radio was planted in the 1890s, when Guglielmo Marconi began his experiments in wireless telegraphy. Soon he was joined by dozens, then hundreds, of others who were enthusiastic about sending and receiving messages through the air — some with a commercial interest, but others solely out of a love for this new communications medium. The United States government began licensing Amateur Radio operators in 1912.

By 1914, there were thousands of Amateur Radio operators — hams — in the United States. Hiram Percy Maxim, a leading Hartford, Connecticut inventor and industrialist, saw the need for an organization to unify this fledgling group of radio experimenters. In May 1914 he founded the American Radio Relay League (ARRL) to meet that need.

ARRL is the national association for Amateur Radio in the US. Today, with approximately 170,000 members, ARRL numbers within its ranks the vast majority of active radio amateurs in the nation and has a proud history of achievement as the standard-bearer in amateur affairs. ARRL's underpinnings as Amateur Radio's witness, partner, and forum are defined by five pillars: Public Service, Advocacy, Education, Technology, and Membership. ARRL is also International Secretariat for the International Amateur Radio Union, which is made up of similar societies in 150 countries around the world.

ARRL's Mission Statement: To advance the art, science, and enjoyment of Amateur Radio.

ARRL's Vision Statement: As the national association for Amateur Radio in the United States, ARRL:

- Supports the awareness and growth of Amateur Radio worldwide;
- Advocates for meaningful access to radio spectrum;
- Strives for every member to get involved, get active, and get on the air;
- Encourages radio experimentation and, through its members, advances radio technology and education; and
- Organizes and trains volunteers to serve their communities by providing public service and emergency communications.

At ARRL headquarters in the Hartford, Connecticut suburb of Newington, the staff helps serve the needs of members. ARRL publishes the monthly journal *QST* and an interactive digital version of *QST*, as well as newsletters and many publications covering all aspects of Amateur Radio. Its headquarters station, W1AW, transmits bulletins of interest to radio amateurs and Morse code practice sessions. ARRL also coordinates an extensive field organization, which includes volunteers who provide technical information and other support services for radio amateurs as well as communications for public service activities. In addition, ARRL represents US radio amateurs to the Federal Communications Commission and other government agencies in the US and abroad.

Membership in ARRL means much more than receiving *QST* each month. In addition to the services already described, ARRL offers membership services on a personal level, such as the Technical Information Service, where members can get answers — by phone, e-mail, or the ARRL website — to all their technical and operating questions.

A bona fide interest in Amateur Radio is the only essential qualification of membership; an Amateur Radio license is not a prerequisite, although full voting membership is granted only to licensed radio amateurs in the US. Full ARRL membership gives you a voice in how the affairs of the organization are governed. ARRL policy is set by a Board of Directors (one from each of 15 Divisions). Each year, one-third of the ARRL Board of Directors stands for election by the full members they represent. The day-to-day operation of ARRL HQ is managed by a Chief Executive Officer and his/her staff.

Join ARRL Today! No matter what aspect of Amateur Radio attracts you, ARRL membership is relevant and important. There would be no Amateur Radio as we know it today were it not for ARRL. We would be happy to welcome you as a member! Join online at **www.arrl.org/join**. For more information about ARRL and answers to any questions you may have about Amateur Radio, write or call:

ARRL — The national association for Amateur Radio®
225 Main Street
Newington CT 06111-1494
Tel: 860-594-0200
FAX: 860-594-0259
e-mail: **hq@arrl.org**
www.arrl.org
Prospective new radio amateurs call (toll-free):
800-32-NEW HAM (800-326-3942)
You can also contact ARRL via e-mail at **newham@arrl.org**
or check out the ARRL website at **www.arrl.org**

Chapter 1

Receiving Antennas are Different!

One of the best-known cardinal doctrines of antenna theory is the concept of *reciprocity*. The gain, directivity, radiation pattern, and electrical impedance of an antenna are the same, whether it's used as a transmitting antenna or receiving antenna. While on the most fundamental level, reciprocity is certainly true; in many cases it is over-applied. It can be easy to overlook the fact that transmitting and receiving antennas perform very different roles in the real world, and are subject to very different requirements and priorities. We will often find that factors well outside the control of the antenna are absolutely *not* reciprocal.

We dedicate considerable space in this book to the non-reciprocal and *bi-refringent* behaviors of the ionosphere, and ways to not only mitigate the effects of these, but also to actually take advantage of them! Since the reciprocal properties of antennas have been covered thoroughly in many other excellent works, we will concentrate in this book on the theory and methods that make receiving antennas *different* from transmitting antennas. We will also present evidence to confirm that nearly every Amateur Radio station can benefit from separate transmitting and receiving antennas in many cases.

It is probably helpful, therefore to modify our definition of *reciprocity* for the remainder of this book. When referring to the immutable reciprocal character of an antenna (even down to the atomic level), we will call

The Limitations of Reciprocity

Transmitting and receiving antennas perform different functions! The well-known Theorem of Reciprocity for antennas is often misunderstood by radio amateurs. While it is true that the fundamental characteristics of antennas apply to both transmission and reception, the requirements and priorities of receiving antennas can be vastly different from those of transmitting antennas. The function of transmitting antennas is to radiate power from the transmitter efficiently, while the function of receiving antennas is to present the best signal-to-noise ratio to the receiver. The focus of this book will be entirely on *receiving* antennas, both active and passive, and their associated circuits. There are relatively few cases where a radio amateur cannot benefit from a separate, well-designed receiving antenna or antenna system.

it reciprocity. When we refer to a complete path between a transmitter and receiver, we will refer to the path as being either *unilateral* or *bilateral*.

Something Old, Something New

The concept that transmitting and receiving antennas are different is not exactly new. In the very early days of Amateur Radio, transmitting and receiving functions were entirely separate entities in the typical ham shack. The nearly universal use of the same antenna for transmitting and receiving roughly coincided with the advent of the *transceiver* in the late 1950s and early 1960s. The convenience and generally good performance of the mechanically steerable beam, primarily in the form of the Yagi-Uda array or cubical quad, rapidly accelerated the adoption of the "transceiving" antenna as the norm during this general time frame. As with many other current standard practices in ham radio, it's easy to forget that they weren't always standard practice. Another prominent example of this is the use of coaxial cable, which only became readily available after World War II. Before then, the use of open wire feed line was the "obvious" way to get power from a transmitter to an antenna.

The Insufficiency of Efficiency

The glaring difference in priorities between transmitting and receiving antennas becomes more...well...*glaring*...when we start looking into the concept of *efficiency*.

When designing an antenna for transmitting, efficiency is generally of paramount importance. Our priority is explicitly to convert as much of our precious RF power into electromagnetic radiation as possible, while dedicating as little of that energy as possible to heating up copper wire (or earthworms). Technically, heat generated in an antenna wire or nearby dirt *is* electromagnetic radiation. It's just not at a frequency that does us much good!

As we will discover, some of the most effective receiving antennas are abysmally poor performers when efficiency alone is considered. In fact, some of the best performing receiving antennas are *utterly unsuitable* for transmitting. (The converse is not generally true; most decent transmitting antennas *will* serve as reasonable, though not necessarily outstanding, receiving antennas.)

What a Receiving Antenna Must Do

Most radio amateurs, even seasoned ones, are surprised to learn just *how feeble* the signals are that they deal with on a regular basis. It really is amazing that radio works at all when we start looking at what the actual numbers are, on a *power* basis. For the sake of this discussion, we'll look only at HF receiving systems, as VHF and UHF receiving is even *more* implausible!

A typical S-meter on a general coverage receiver or HF transceiver is calibrated so that a 50 µV (microvolt) signal at the antenna input terminals results in a reading of S-9. This is an old, old standard which is still very useful, as we shall see. A typical modern receiver has a nominal input impedance of 50 Ω. Let's look at the *power* level of an S-9 signal. Now, note that an S-9 signal is a *strong* signal…not something we have to dig out of the noise. How do we figure out the power? Well, since we know the input impedance and the input voltage, we can use E squared over R (E^2/R). Some simple algebra yields the answer: 50 µV is 0.00005 V. We square that, and we come up with 0.0000000025. We now divide that by 50, our load impedance, and come up with the staggering figure of 0.00000000005 W. That's 50 *picowatts*. Or fifty trillionths of a watt. This is for a *strong* signal!

How about an S-1 signal, one that's right near the noise floor? Well, the S-meter scale is designed so that each S-unit represents a 2-to-1 change in voltage, or a 4-to-1 change in power. There are 8 S-units difference between an S-1 and an S-9 signal. If each S-unit is 4-to-1 power ratio, the total power ratio between S-1 and S-9 is 4 to the 8th power (4^8) or 65,536. So to figure out our *power* at S-1, we divide our 50 picowatts by 65,536, which comes out to 0.00000000000000076 W … or 0.76 *femtowatts*! Or 0.76 quadrillionths of a watt. That's not a lot of watts! (By the way, these numbers don't sound anywhere near as impressive…or daunting…when expressed in decibels, the topic of our entire second chapter).

Despite the incredibly minuscule amount of power represented here, we routinely deal with these signals with extremely pedestrian equipment. There isn't a receiver built today that can't create a usable audio output with an S-1 signal! In fact, the lowly regenerative receiver of nearly 100 years ago could do this. You don't need to be impressed with our technology; you just need to be impressed with the physics that allows this all to happen.

Capture This

When it comes to receiving antennas, one very telling measuring stick is what is known as "capture area." While there are no universally accepted units for capture area (you won't find the figure listed on any commercially manufactured ham antennas), the concept is quite useful.

At any appreciable distance from a transmitting source, a receiving antenna can only intercept a minuscule fraction of the total radiated power. As the signal retreats from the transmitting antenna, it spreads out over a large volume of space. Regardless of how much antenna gain exists at the transmitter, the total energy is distributed pretty much spherically. For all practical purposes, we can consider the wavefront from any distant transmitter as being spherical. This means that the signal intensity will diminish as the inverse square of distance.

A receiving antenna can only intercept that portion of a signal that would ordinarily pass through its "personal space." There is nothing in a receiving antenna that can "pull" a radio wave toward itself, contrary to some advertising literature to the contrary. Sales terms such as "Wave Magnet" or "Signal Grabber" were fairly commonly employed in the past by manufacturers touting the capabilities of their receivers and their associated antennas. Fortunately, we don't hear too much of this language any more. A radio wave goes where it goes, and if a receiving antenna happens to be in the wave's predetermined path, it can extract energy from the wave. If not, it can't. If you could actually invent a wave magnet, you could become very rich. In reality, the only way you can grab more signal is with a bigger mitt…or antenna. The more "sky" your receiving antenna occupies the more radio wave it can intercept.

Now, there is a bit of a subtlety we need to add to this concept. Because of the complex interaction of the electric *and* magnetic field of an electromagnetic wave (including the inevitable *re*-radiation from any receiving antenna), a wire antenna actually has a bit more capture area than the actual cross sectional area of the wire exposed to the passing wave. To fully explain this would require delving into Maxwell's equations a bit more deeply than most of us would care to, or need to…at least for the purposes of this book. From another vantage point, capture area for an antenna is approximately equivalent to "aperture" as applied to a lens, at optical wavelengths.

The salient point here is that the more capture area you have, the stronger your received signal will be. All other things being equal, it's fair to say that the longer the antenna, the better it receives. This is consistent with the concept of Total Copper (or aluminum) Content (TCC) as a yardstick of overall Amateur Radio station performance. The more copper (or aluminum) you have in the air the better you'll talk and the better you'll hear!

Diminishing Returns

Taking the previous discussion at face value, it would appear that the gain of a receiving antenna will increase indefinitely as a function of the aperture, to the point where the aperture is so large that it intercepts *all* the transmitted energy. Of course, the only time this would happen would be if you entirely surrounded the transmitter antenna with your receiving antenna — not a very practical solution. (It is also illuminating to go the other way. Shrink the length of a dipole to where is a very tiny fraction of a wavelength, and notice that the aperture area limits to $l^2/4\pi$, and directive gain never gets below 1.5 or 1.76 dB compared with 1.64 or 2.15 dB for a half wave dipole.)

For a simple single-wire antenna, the gain advantage versus length reaches the point of diminishing returns on the order of a half wavelength or so. The 5/8 wave vertical or the "extended double Zepp," at twice this length, generally represent the maximum gain scenario for single-element antennas. One notable exception to this constraint is the *wave antenna,* such as the Beverage, where, indeed, the gain continues indefinitely as the length is increased. We will discuss wave antennas in detail in later chapters.

Signal-to-Noise, the Bottom Line

There is a strong temptation to evaluate the performance of a receiving antenna by looking at the S-meter alone. Although the gain of a receiving antenna is indeed reflected in the S-meter reading, the connection between readability and S-meter reading can be quite indirect at best, and utterly meaningless at worst. The ultimate goal of a receiving system isn't to exercise the springs on your S-meter's D'Arsonval movement (or at least it shouldn't be), but to optimize intelligibility, which is a function of signal-to-noise ratio than actual gain. As we will discover, it is often to great advantage to sacrifice a little bit of raw gain in order to achieve a much greater reduction in noise — and thus significantly increase the signal-to-noise ratio at the receiver's input. That's our ultimate goal. In fact, the bulk of this book is actually centered on devices and methods specifically designed to do just that.

What is Noise?

In order to achieve our supposed goal of optimizing signal-to-noise ratio, it's obviously useful to know what noise is. In simplest terms, noise is what we don't want to hear, and signal is what we do want to hear. Signal is useful, meaningful information. It is what we generate on the transmitter side of the equation. Noise is all the accumulated non-signal stuff that arrives at your antenna. Interference is generally *not* included in the definition of noise, although it can be. Effective receiving antennas can reduce or eliminate certain types of interference, as well as reduce noise.

The noise we will primarily deal with is either natural ionospheric and atmospheric noise ("static"), or manmade electrical noise, generally more localized in nature. While receiver or "electronic" noise is also commonly included in the definition, it is generally not a factor in modern well designed receivers at HF and lower frequencies. We'll learn why this is so in short order.

Chapter 2
Your Friend, the Decibel

Contrary to popular Amateur Radio lore, the decibel was not created to make life difficult. As we will discover, the decibel can greatly simplify a number of antenna-related measurement issues. Although you can probably survive your entire Amateur Radio career without ever learning how to work with decibels, there is really no need to.

Like all modern electrical units officially accepted as SI (International System) units, the *decibel* is always spelled in lower case. However, unlike most other SI electrical units, which are named after dead science guys, the decibel is only *partially* named after a dead science guy (Alexander Graham Bell), and was only relatively recently formally accepted by the SI committee. There is more information about SI units than you'd ever want to know, right here: **www.nist.gov/sites/default/files/documents/pml/div684/fcdc/sp330-2.pdf**

The decibel allows you to make simple, meaningful comparisons of antennas and other signal path components. As we discussed in the previous chapter, we sometimes work with minuscule numbers, which have lots of zeroes after the decimal place. We can also work with some fairly large numbers, with a lot of zeroes in *front* of the decimal place (though we will deal with relatively few of these when talking about receiving methods). In any case, we will be talking about *very* large *ratios* of voltages and powers, which can be rather inconvenient and daunting to work with using "normal" methods. The decibel is wonderful at compressing very large power ratios into bite-sized chunks of arithmetic. And, furthermore, the decibel can convert many complicated multiplication problems into simple addition problems. So, if the decibel isn't second nature to you, it's very much worth your time to make it so. It really is your friend.

Power is King

The decibel tells us the power ratio between two locations in a circuit, or sometimes the same location in a circuit at different times. While

Power is Power

In Amateur Radio literature, occasionally you may run across the terms "Power Decibels" and "Voltage Decibels." This can be very misleading. Decibels are *always* related to power. There *is* such a thing as the *voltage-derived* decibel, as well as the *current-derived* decibel, but the answer is still the same as long as the input and output impedances are identical. The decibel is always a *power* ratio. The bottom line or goal of any antenna system (or communication system, for that matter) is to optimize the power transfer from point A to point B. Many different proportions of voltage and current may be used to achieve this end, but it is always *power* that does the work.

It is absolutely crucial, when using voltage-derived decibels, that the *impedance* across which the voltage is measured is constant. Otherwise you could make the case that a simple transformer can give you power gain...which, of course, is physically impossible. The law of NFL (No Free Lunch) tells us we can never get more power out of a device than we put in. And most of the time, we're pretty lucky if we can even achieve that!

---

the figures we use to *calculate* decibels may be in either voltage, current, or power, the final answer in decibels is always a power ratio. Regardless of how we go about it, the end result in any radio communications circuit is optimizing the *power* arriving at the receiving end. And, as such, the decibel is ideally suited to evaluating just how well we're achieving that end.

The formula for decibels (dB) is:

$$dB = 10 \log_{10}(P1/P2)$$

where P1 is the power coming out of a device, and P2 is the power going into the device.

The final answer can be a positive or negative number, depending on whether we have a signal gain, or a signal loss.

For 99.999% of all we do in Amateur Radio, we will use the common log, or base-10 logarithm for all our decibel calculations. There is another form of the decibel, called the natural log decibel, or the *neper*, from *Naperian* logarithms. The neper is sometimes used by microwave and space communications engineers, but is becoming somewhat less common even among these groups. This will be the last mention of the neper we will make in this text.

Indirect

One of the odd paradoxes we encounter in this discussion is that, while power is our end goal, we almost never measure power directly. Apart from some rarely used (at least in ham radio) *calorimetry* methods, we must derive power measurements indirectly, either by measuring voltage or current, and then applying E^2/R or I^2R to arrive at a power figure at any given location. Furthermore, while RF current measurements are sometimes used in Amateur Radio transmission systems, by far the most common *direct* radio frequency measurements are *voltage* measurements. RF voltage probes, oscilloscopes, and even the venerable Bird directional watt meter are actually voltage instruments.

Because of the difficulty of measuring power directly, the *voltage-derived decibel* is commonly used. The formula for the voltage derived decibel is:

$$dB = 20 \log_{10}(V1/V2)$$

where V1 is the output voltage from some device, and V2 is the input voltage into the device. It can be positive or negative in this case too.

Danger!

When using the voltage derived decibel we *must* be very careful that the voltages for V1 and V2 are measured across exactly the same resistance (or impedance)! Decibel figures will be meaningless if this rule is not strictly adhered to. We will shortly demonstrate an extreme example of how violating this rule can yield impossible results. Many hams fall into this trap when working with dB. But if you keep reminding yourself that dB is always a *power* ratio, you can (usually) avoid the pitfall.

In normal amateur transmission systems, where transmission lines and transmitter outputs are (at least nominally) 50 Ω, we can use our voltage derived $dB = 20 \log_{10}(V1/V2)$ formula. For example, measuring the RF voltage at the input and output of a *flat* (perfectly matched) transmission line will give us the information we need to calculate the dB of attenuation (loss) through the transmission line. We are also fairly safe if we want to measure the dB gain through a "brick" type modular amplifier, where the input and output impedances are intentionally tweaked to be exactly 50 Ω. However, many of the antennas and amplifiers we will be dealing with in this book will have *drastically* different input and output impedances, so we cannot use this voltage derived dB. As a general rule, we will assume that this book is "not safe for $dB = 20 \log_{10}(V1/V2)$. We will explicitly note when it *is* safe to use this formula.

A Happy Exception

One instance where it *is* nearly always safe to use voltage derived dB is when we're taking measurements across the *exact same point* in a circuit, such as across the input terminals of a receiver with two different input levels. This is why an S-meter can give us reasonable dB comparisons, because we're comparing the exact same terminals but at different times. Fortunately this exception to the "not safe" rule is a very common one, and we will use it with confidence, when appropriate.

Insanity Check

Let's explore a misuse of the voltage-derived dB formula and show why it is so crucial to only use this across the same impedance.

Let's take the case of a common 10:1 voltage step-up transformer (**Figure 2.1**). From our elementary electronics, we know that this transformer has a 10:1 turns ratio; there are 10 times the number of turns on the output as on the input. Just to keep things tidy, let's assume our transformer is 100% efficient; there are no losses in the transformer itself. Whatever power we put into the primary is available on the secondary... no more, no less. To keep the number simple, let's put 1 turn on the primary, and 10 turns on the secondary. If we put 10 V ac across the primary, we know we're going to get 100 V across the secondary. Nothing too mysterious here.

Now, we don't ordinarily just leave the output terminals of a transformer flapping in the breeze. We want to do some actual work with our transformer, so let's put a load resistor on the secondary. We'll make our load resistor a nice round 1000 Ω. Okay, so let's figure out the power dissipated in our resistor. The first thing we might want to do is figure out the current, which is simply E/R or 100/1000, or 0.1 A. We can then use our

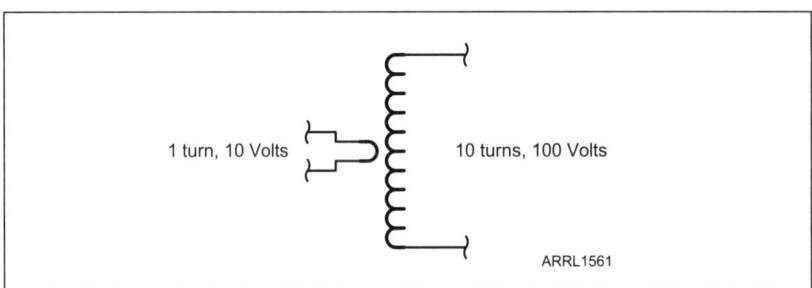

Figure 2.1 — A transformer with 1 turn on the primary and 10 turns on the secondary has a 10:1 turns ratio (there are 10 times the number of turns on the output as on the input).

simple power formula, P = I × E, which gives us 0.1 × 100 or 10 W.

Since our transformer is neither magical, nor abysmal, we know that we have to put 10 W *into* the transformer to get 10 W out. Using our simple P = I × E, we can divide P by our known input voltage (E), and come out with 1 A. The product of the input voltage and current has to equal the product of the output voltage and current. No rocket science yet.

Now, let's figure out our transformer gain (or loss) in dB. Using our original basic formula, dB = 10 $\log_{10}$ (P1/P2), we find that P1/P2 is 1. The $\log_{10}$ of 1 is 0, and 10 × 0 is still 0, so our gain is 0 dB. We've neither gained nor lost anything. (We are fully aware that there is no such thing as a 100% efficient transformer, but we'll make an exception for this example.)

Let us now show the gross error that the voltage derived dB will yield in this exact same circuit. Let's measure the output voltage: 100 V. This will be our V1. Let's measure the input voltage: 10 V. This will be our V2. Now let's plug those values into our voltage derived formula, dB = 20 $\log_{10}$ (V1/V2). V1/V2 = 10. The $\log_{10}$ of 10 = 1, so 20 × 1 = 20. Final answer is 20 dB. Our transformer is amazing! It just increased our power by 20 dB! That's a 100-fold increase in power, all by magic. We can sell this transformer online and be rich rich rich! Or not.

What is the problem?

The problem is the input and output impedances of the transformer have not been taken into account! In reality, the impedance ratio of an ideal transformer is the turns ratio *squared*.

Our transformer has a turns ratio of 10:1, which gives it an *impedance* ratio of 100:1. For the sake of argument, we'll assume the output impedance of our transformer is matched to the 1000 Ω resistor. Using the turns-squared principle, we now know that the input impedance is 10 Ω.

Well, just as a sanity check, let's work out the power using E^2/R, now that we know what the R values are. (In a perfect transformer, these R values will also be purely resistive, while in the real world, they also have leakage inductance). For the input $E^2/R = 10^2/10$, which is 10 W. On the output we have $100^2/1000$, which is 10,000/1000, which is also 10 W. All is now right in the universe!

While the misuse of the voltage-derived dB can give us inflated ideas about the performance of our antenna or other devices, it can also give you an overly pessimistic view. And, in the case of active antennas, erring on this pessimistic side is far more frequent. We will demonstrate clearly how a device such as a *voltage follower* amplifier can yield nearly infinite power gain, in real terms, while using the voltage derived formula will actually tell you that there's a small power *loss*. Since the voltage follower is such an integral part of active antennas, we want to reveal all the stumbling blocks to meaningful measurements right up front. We will discuss this topic many more times in great detail as we progress.

Perhaps now is a good time to demonstrate the real value of dB when it comes to complex radio circuits and paths.

A Comforting Anchor

Recall that the dB is a measurement of power *ratios*. Since the bulk of what we do in radio involves *linear* circuits, that is circuits and components that produce an output that is proportional to an input, regardless of the *absolute* value of that input, the "pure" decibel is an extremely handy device…and a great reflector of physical reality. There are times, however, when we want to deal with *absolute* levels of power, and yet be able to employ the flexibility and convenience of the decibel. What we want in this situation is the *referenced* decibel, or what I like to call the *anchored* decibel. The referenced decibel is a decibel with a "starting point" of an absolute fixed value of power. For a number of reasons, the *milliwatt*, or 1000th of a watt, is a very convenient anchor point for our decibels. We'll do a fun and revealing exercise to show the brilliance of the decibel, and trust that after this, you will truly appreciate your friend the decibel.

Recall in Chapter 1 the requirements needed to receive a typical HF signal. We'll convert some of those numbers into decibels, and add a few tweaks, which will give you a good "feel" for what we need to accomplish.

dB Phone Home

Many decades ago, "Ma Bell" developed some standards that we still use today. The nominal audio level for a comfortable listening level on a standard land line (remember those?) telephone handset was determined to be about 1/20th of a milliwatt (mW), or 1/20,000th of a watt. That's not very much power. This would be the equivalent of an audio "S-9" signal. But let's figure out how much gain in dB we need to achieve that S-9 signal at the headphones, with our 50 µV antenna signal.

Well, to allow us to use decibels, we really need to drop our anchor. And let's drop it at precisely 1 mW. We'll call 1 mW 0 dBm. This is a decibel referred to 1 mW. When working with anchored dB, our formula is exactly the same, but we nail our P2 value down at an absolute value, in this case, 1 mW.

Our first task is to determine, in dBm, how much power our 1/20 mW is. Recall that it is our P2 value that is anchored, and it's anchored at 1 mW. P1 is 1/20 or 0.05:

$$dBm = 10 \log_{10}(0.05/1)$$

Taking the log of 0.05/1, we get −1.3. Multiplying that by 10 we get −13. −13 *what* you ask? It's −13 dBm. The minus indicates it's *less* than 0 dBm, or 1 mW. So, in Ma Bell terms, the standard phone audio level is −13 dBm.

Now we have a target to shoot for. We need to bring our 50 µV radio signal up to −13 dBm to have a comfortable listening level, with a "normal" pair of headphones. How much overall gain will our receiver need to have to achieve this?

First, let's convert our input power into dBm, to keep everything on the same page. Remember that 1 mW and 0 dBm are the same values. Recall that 50 µV across a 50 Ω load gives us 50 picowatts. How many dBm is 50 picowatts?

The simplest way to do this is to simply scoot the 50 picowatts three decimal places to the left, since a milliwatt is three decimal places to the *right* of a watt. This gives us a value, in milliwatts of 0.00000005. Plugging this into our dBm formula, we find that the log of 0.00000005/1 is −7.3 and some spare change. Multiplying that by 10, we get the grand total of −73 dBm.

So our original 50 µV at our antenna terminals translates to −73 dBm. We need to bring that level up to the −13 dBm value we need in our headphones. This means we need to come up with 60 dB of gain, to make up the difference. 60 dB is a 1,000,000:1 (one million to one) power ratio, which really isn't all that hard to come up with. You can easily do it with three transistor stages. The nice thing about this is that you can either use RF gain, or IF gain, or audio gain, or any combination to come up with that extra 60 dB. In the early days of radio, most hams opted for gobs of RF gain, using devices such as regenerative detectors, and just enough audio gain to drive a sensitive headset. The trend nowadays is to move the gain farther down the "food chain." Scientific grade direct-conversion receivers use just enough RF gain before the mixer to establish the noise figure, and use tons of post-detection "audio" gain to do the rest. The latter method can yield superior dynamic range, when well-implemented.

Now, keep in mind that the above figures are for a strong S-9 signal. What do we need for signals down near the noise floor, or S-1? Recall from the previous chapter that there is a power ratio of 65,536 between an S-1 and an S-9 signal, assuming 6 dB (4-to-1) change per S-unit. How many dB is 65,536:1? Using our standard dB formula, we come out with 48 dB and a few cents change. So an S-1 signal is 48 dB below our −73 dBm figure, which comes out to be −121 dBm. To bring an S-1 signal up to normal listening level requires another (approximately) 100,000 times power gain, which requires another three stages of gain, typically. As a sanity check, you can look at the spec sheet for a typical HF ham transceiver, and they will give a "sensitivity" figure of typically –120 dBm or so.

Again, as demanding as these figures seem to be, they are easily achievable with off-the-shelf equipment. My 1940 vintage Hallicrafters Super Defiant receiver can do this…at least on the low bands!

As you have seen, the decibel can make short work out of some unwieldy numbers. There is no reason to avoid using this valuable tool.

Chapter 3

The Preamplifier Problem

RF gain is incredibly easy to achieve in bulk quantities, and therein lies a big problem. In theory, one can cascade any number of medium-gain RF amplifiers together and thus approach infinite gain and sensitivity. If the end goal is achieving high S-meter readings and massaging the ego of the ham on the transmitting end, the promiscuous application of preamplifiers will fill the bill admirably. If the goal is increasing the true sensitivity of your receiver, under less-than-admirable conditions, the typical RF preamplifier is about the last thing you need or want.

The truth of the matter is that the vast majority of HF receivers on the market today have more RF gain than necessary...or even usable. While it was true that some lower end "boat anchor" receivers could benefit from outboard preamplifiers, especially on the higher bands, this hasn't been the case in decades.

It should be obvious (though evidently it isn't) that a preamplifier ahead of the "front end" of an HF receiver will amplify noise just as much as the desired signal. The absolute limiting factor for usable RF sensitivity is *thermal agitation noise* of the antenna. If your receiver is capable of detecting thermal agitation noise, no amount of gain *after* said antenna can do you any good whatsoever.

More than Enough

When they are used with a full-sized antenna, nearly all, if not all, modern HF receivers have far more than enough gain for effective HF operation without a front-end preamplifier. The use of such preamplifiers often causes more harm than good. We will demonstrate how the sensitivity of a receiving system at HF (frequencies below 30 MHz) is limited entirely by *thermal agitation* noise, which a typical preamplifier boosts just as much as the desired signal. A high-performance *unity gain* preselector is an alternative to the typical broadband preamp. Signal-to-noise ratio is far more important than received signal strength in most situations.

The Screwdriver Test

In **Figure 3.1** you will see my Highly Technical Test Instrument (HTTI). This sophisticated instrument will reveal whether your receiver can benefit from the use of an RF preamplifier. The instrument consists of a short screwdriver with a banana plug, to assure a reliable contact with your receiver's UHF (SO-239) antenna input connector. Tune your receiver or transceiver to the high end of 10 meters. Set your IF bandwidth to the widest bandwidth available. Disconnect whatever antenna you may have. Turn up your RF gain to the maximum, and the audio to a level where you can hear the receiver's internal white noise. Now insert the Highly Technical Test Instrument into your antenna connector. If you hear an increase in white noise...to any degree...your receiver has all the RF gain it can possibly use.

What Am I Hearing?

The noise you hear from the Highly Technical Test Instrument is a combination of *thermal agitation noise* and *cosmic noise.* Thermal agitation noise is the noise the steel of the screwdriver generates by the mere virtue of its existence in a non-absolute-zero-temperature environment. Cosmic noise is the wideband background noise of the universe, which can be surprisingly large. We will discuss this noise source in a subsequent discussion of the *riometer.*

These two noise sources determine the absolute limit of your *useful* receiver sensitivity. Any additional preamplification will increase this noise along with any signal. The only way you can decrease thermal agitation noise is to soak the screwdriver in liquid helium before plugging it into your receiver. The only way you can reduce *cosmic* noise is by inserting the Earth between your antenna and the center of the universe. (This is why cosmic noise undergoes very predictable *diurnal* variations.) Whether you're using a screwdriver, a coat hanger, or a long-wire antenna, it's the thermal agitation noise and cosmic noise that set the limit on usable receiver RF gain and sensitivity.

While thermal agitation noise and cosmic noise are the *limiting* noise factors, they are by no means the only noise factors. Atmospheric noise is generally significantly greater than the thermal agitation and cosmic noise, except at the upper regions of HF. This is why for the screwdriver test we tuned to the upper end of 10 meters, where atmospheric noise is essentially nonexistent.

Figure 3.1 — The Highly Technical Test Instrument (HTTI).

But What About VHF/UHF?

Thermal agitation noise certainly exists at VHF and UHF frequencies. In fact, thermal agitation noise is a highly "white" noise, basically covering the radio spectrum from dc to daylight. However, at VHF and UHF (and above), the internal noise of the receiver's active devices become major contributors to noise, usually far exceeding the thermal agitation noise. In this case the limiting factor for sensitivity is the device noise. Since the noise figure of any array of cascaded amplifiers is determined by the first stage, it is important that the front end contributes as little noise as possible. VHF and UHF systems can thus benefit from the use of low-noise preamplifiers, especially if they are installed at the antenna, ahead of any lossy (and noise producing) transmission lines.

A Viable Alternative

The indiscriminate use of wideband RF preamplifiers has at least three undesirable effects.

1) Increase in overall system noise. Since the front end of any receiving system is determined by the first stage, if an RF preamplifier does not have exceptionally low noise itself, any noise it contributes will be amplified throughout the receiver chain. In the best case scenario, an RF preamplifier will contribute no additional noise.

2) Overloading by strong signals. Preserving the dynamic range of a receiver requires that no overloading occurs anywhere along the path. Signals that would not normally overload the IF or other stages of a receiver very well could do so with the indiscriminate use of a high-gain RF preamplifier.

3) S-meter "inflation." While a few transceivers compensate the S-meter sensitivity when the preamplifier is engaged, most do not. A well calibrated S-meter is one of the most underrated accessories for effective Amateur Radio operation. Of course, the most precise S-meter won't do much good if the operator doesn't pay any attention to it. Amateur Radio progress, oddly enough, relies on accurate, believable signal reporting, among other things. As we will explore later on, the S-meter has gotten somewhat of a bad rap. Although the instrument as deployed on many earlier receivers has been notoriously inconsistent, modern receivers have more trustworthy S-meters.

If preamplifiers are so counterproductive, is there a good alternative? Indeed there is: the high-Q, unity gain preselector. A unity gain preselector will not compromise the dynamic range of a properly designed receiver, but rather can assist the dynamic range by restricting the number and intensity of nearby signals that could cause intermodulation or other related issues. Unlike the front end Q-multipliers of days gone by, which greatly increased selectivity *and* gain, in a sometimes counterproductive fashion, the unity gain preselector maintains the same gain regardless of the Q, which can be extremely high. Modern circuit topologies allow this behavior with considerable simplicity.

In the next chapter, we will explore the *good* news when it comes to really small receiving antennas.

Chapter 4

The Amazing Disappearing Antenna

First, the Bad and the Ugly

One of the immutable facts of life when it comes to Amateur Radio is that efficient HF transmitting antennas are necessarily large. While one's *absolute* definition of large may vary, the fact remains that unless an antenna is a significant percentage of a wavelength long, either its efficiency or its usable bandwidth will be compromised, often severely. This doesn't mean that very short transmitting antennas can't work. The ARRL 600 Meter Experimental Group (**www.500kc.com**) has shown just what can be done with vertical radiators between $\frac{1}{50}$ and $\frac{1}{20}$ of a wavelength. But it takes great attention to detail and out-of-the-ordinary skill and methodology to receive signals at any distance from these highly abbreviated transmitting antennas.

In recent years, hams seeking that elusive high-efficiency miniaturized HF transmitting antenna have been the prime targets of a number of Amateur Radio snake oil salesmen. Some of these "wonder" antennas rely entirely on transmission line radiation for their functioning. This can be easily verified by removing the transmission line entirely and inserting a small battery powered transmitter right at the feed point, and then performing a rudimentary field strength measurement. With a small class of notable exceptions, these antennas fail this test miserably.

The notable exceptions are the relatively efficient *small tuned loops* or STLs. My latest pilgrimage to the Dayton Hamvention was eye-opening, in this regard. At least half a dozen manufacturers of high-quality STLs were represented. All of these antennas were constructed of large diameter tubing, which is the only way to significantly minimize the losses of these otherwise "small" antennas. The total copper content (or aluminum content) of these antennas is significant. But these STLs are the

The Active Antenna

The "deficiencies" of very short or small receiving antennas are not deficiencies at all, but can be used to great advantage. The aperiodic (untuned) active antenna is a prime example of an extremely small antenna having outstanding performance...if designed properly. Very recent developments in op-amp technology make the design of aperiodic active antennas simpler than ever.

exceptions that prove the rule that efficient transmitting antennas are generally a significant fraction of a wavelength in size.

This in no way implies that if you can't put up a full sized or nearly full sized antenna you are out of luck. There are *proven*, reliable means of increasing the effectiveness of very short HF antennas, none of which require any proprietary black magic. You need look no farther than the countless hams who have very effective 80 meter, and even 160 meter mobile stations.

And Now for the Good

Fortunately, size doesn't matter so much when it comes to receiving. In fact there are a number of reasons why a very small antenna (in terms of wavelength) may be preferable when it comes to receiving effectiveness. While small receiving antennas have the obvious advantage of being able to fit neatly into acreage-restricted antenna farms, this is by no means the only advantage. Here are just a few more:

1) Radio direction finding, either mobile or on foot. While you may not feel particularly motivated to go dashing through the snow with hordes of avid fox hunters just for the fun of it (though most of us old timers could certainly benefit from the exercise), you will most likely encounter at some point in your ham career the need to track down an offending signal of some sort. This would be a rather daunting task if receiving antennas couldn't be made small!

2) Noise or interference nulling. Small antennas of most types tend to have very sharp nulls in certain directions, generally much sharper and deeper than full-sized antennas. These very deep nulls can be effective in eliminating noise or interference from single point sources.

3) Easily customized beam patterns. Arrays of small receiving antennas allow some very interesting possibilities for both "normal" amateur operation as well as radio science....even if you *do* have plenty of room for full-sized antennas.

4) They can be really cute. My eXOgon antenna, which I shall describe in full in a chapter near the end of this book, makes an absolutely intriguing lawn ornament. If you live in a deed restricted neighborhood of some kind, the president of your homeowners association (HOA) may actually *request* that you install a few of these. Well, probably not, but hope springs eternal. (We are currently developing another receiving antenna cleverly disguised as a plastic flamingo.)

In addition to "very" small receiving antennas, there are numerous very effective "semi-small" receiving antennas, which we will describe in detail. Properly installed "very small" and "semi-small" antennas can often exceed the performance of full sized antennas. And it can even be argued that the Beverage antenna, which is generally very *long*, can be considered a "small" antenna, in terms of obtrusiveness, visual or otherwise.

Distance, the Great Equalizer

As we demonstrated with our Highly Technical Test Instrument (HTTI) screwdriver antenna (not to be confused with the trademarked and capitalized Screwdriver mobile Amateur Radio antenna), it doesn't take much antenna to generate some thermal agitation noise. It *does* take a little bit of receiver potential to hear the phenomenon, but nothing out of the ordinary. But how much antenna do we need to pick up an actual usable *signal?* The surprising answer is, not much more! I have a little $40 Grundig shortwave pocket radio, and it has a little telescoping whip. When there is good propagation, I can hear all kinds of shortwave broadcast stations with the little antenna fully retracted. When there is bad propagation, I can't hear much even with the whip fully extended. But I can't hear anything on my full sized inverted V, either.

The dynamic range (the ratio of path loss between excellent and no propagation) of the ionosphere is essentially infinite. The dynamic range between the HTTI screwdriver and any *practical* high gain HF Yagi is about 25 dB....maybe. Radio propagation is the great equalizer, the difference between the best and the worst receiving antenna pales in comparison with ionospheric variables. And even when propagation behaves itself, the free space attenuation of *any* radio signal is, after you move a few wavelengths from the source, of far greater variability than the difference between a good receiving antenna and a lousy one. Remember that, in free space, the field intensity of a radio wave decreases as the square of the distance. In the "real world" the attenuation is generally greater, often much, much greater.

The Character of a Very Short Antenna

Although not all active receiving antennas are extremely small (in terms of wavelength), the very short whip is interesting enough to merit its very own chapter. This is certainly not to ignore the extremely important loop antenna, which we will explore in depth as well. However, the "linear" antenna is a bit easier to analyze than the loop so we will begin with that. As we delve into the surprising characteristics of a very short whip, be prepared to jettison some of the common wisdom you may have accumulated about antennas.

As a baseline, let's go ahead and model a full-sized ¼ wave whip antenna over a perfect, infinite ground plane. This will serve as a reference antenna for comparison with our magical disappearing whip. It should be understood that you're never going to encounter a perfect, infinite ground plane, but as with many other references we use in antenna work, it's a *very* useful tool. We can do this for any frequency, but 40 meters (exactly 7 MHz) gives us some nice round numbers to work with. As with all the models in this book, we will use *4nec2*, but you will get the same results with any antenna modeling program.

The geometry of this antenna is shown in **Figure 4.1**. We have "cut" the antenna to precisely ¼ wave at 7 MHz. We have inserted a voltage source right at the bottom of the antenna, just above ground. And we have selected "perfect ground" from the pull-down menu of ground types. And then we've run a far field analysis.

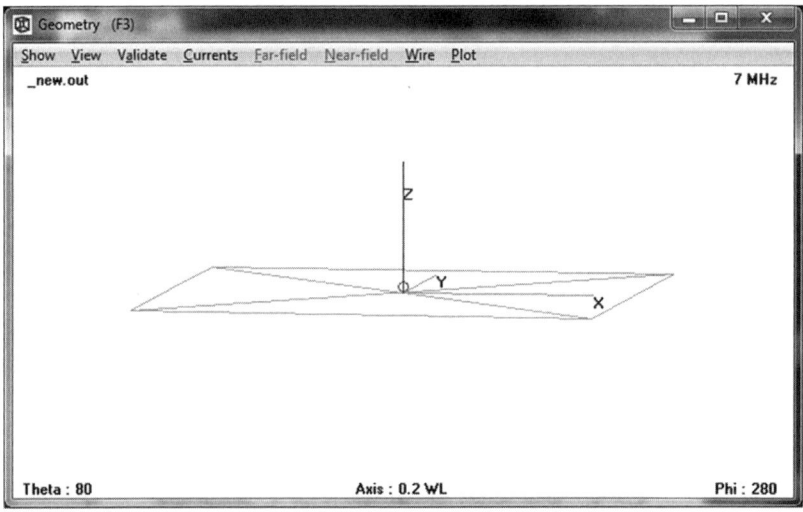

Figure 4.1 — Model of a ¼ wave vertical antenna over a perfect ground.

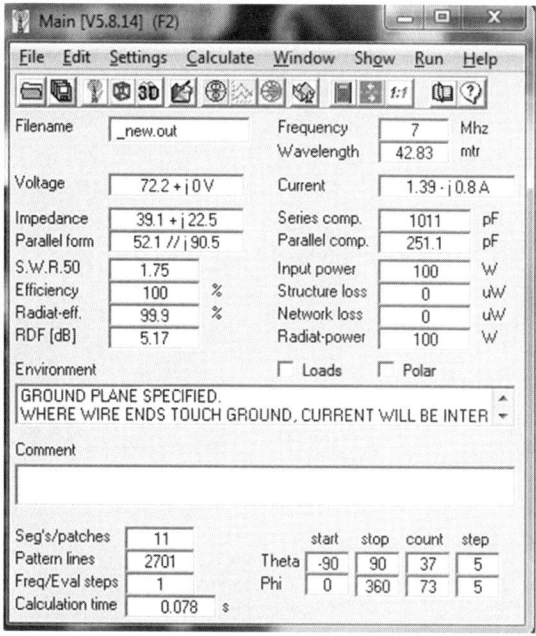

Figure 4.2 — Modeled impedance data for the ¼ wave vertical antenna of Figure 4.1.

After the model is run, we now have the impedance information is shown in **Figure 4.2**. And finally, the radiated pattern is shown in sparkling 3D in **Figure 4.3**. None of this should be too surprising for any experienced ham. Inspecting Figure 4.2, we see that the input impedance is 39.1 + j22.5 Ω. Well, there might be a *tiny* surprise here. We've always been told that a ¼ wave vertical over perfect ground is 36 Ω…and perfectly resonant. However, this is assuming a perfectly thin wire. Since the wire is of finite diameter, there's a bit of capacitive loading "end-effect" afforded, which gives us a small amount of effective lengthening. That accounts for the j22.5 Ω. And, the radiation resistance is just a tad above the theoretical 36 Ω, again because the antenna is effectively *slightly* longer than ¼ wave, electrically. We can twiddle either the

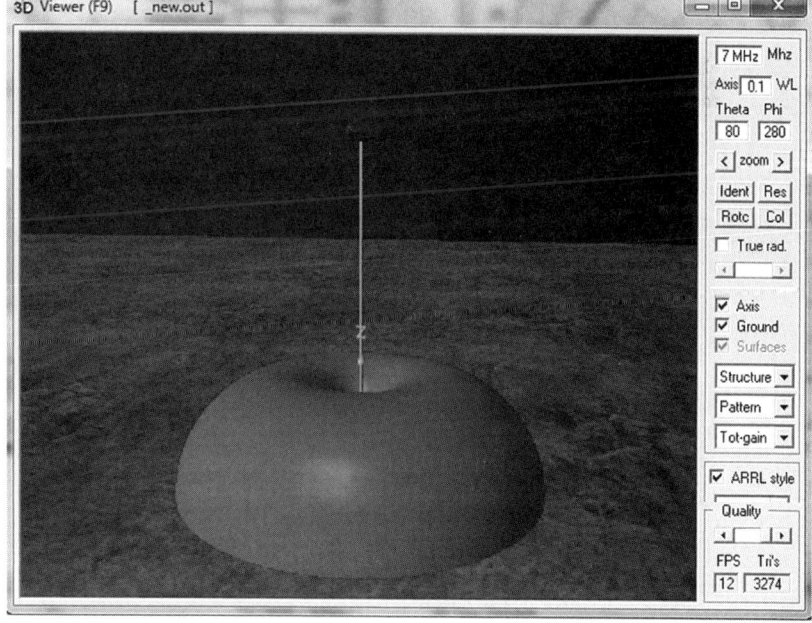

Figure 4.3 — 3D radiation pattern of the ¼ wave vertical of Figure 4.1.

The Amazing Disappearing Antenna   4.5

frequency or the length to give us our 36 Ω, perfectly resonant condition, but it won't really help us understand the topic at hand.

Now, there isn't really much to say about the radiation pattern, except that it's half of your classic dipole "doughnut" pattern…again no big surprise. The antenna is omnidirectional in azimuth, with a well-defined "cone of silence" overhead.

Now, let's go ahead and shrink the antenna to $1/10$ of its current size, making it $1/40$ of a wavelength. This would be fairly representative of the collapsible whip antenna on an el cheapo shortwave radio receiver... with the exception that you won't have a perfect ground plane on your el cheapo shortwave radio!

The geometry is shown in **Figure 4.4**, the impedance information is in **Figure 4.5**, and the radiation pattern is in **Figure 4.6**. Well, what have we here? The impedance is drastically different now, isn't it? We now have $0.27 - j2410$ Ω. The radiation resistance is basically nonexistent,

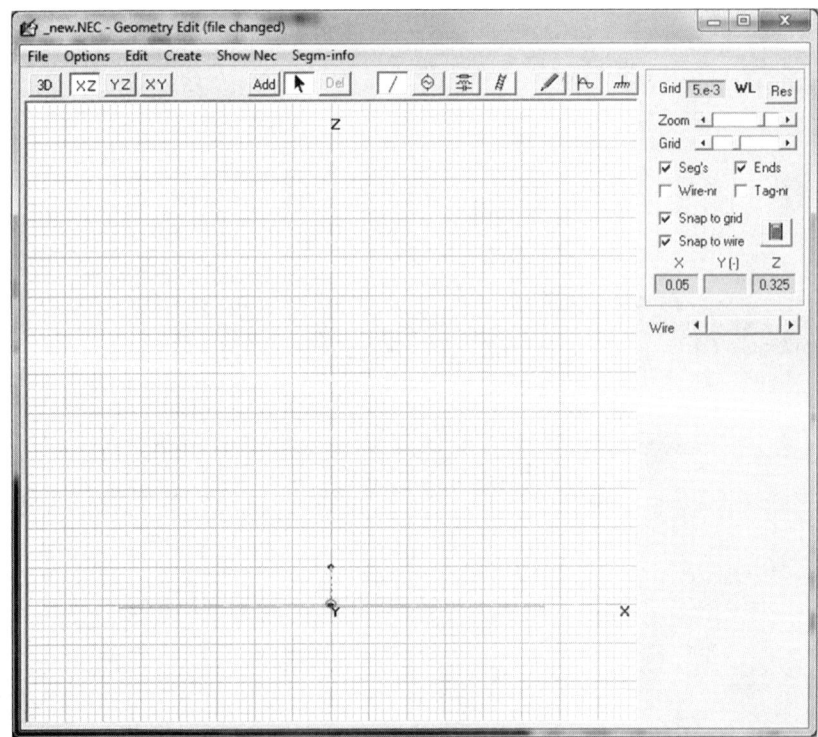

Figure 4.4 — Model of a $1/40$ wave vertical antenna over a perfect ground ($1/10$ the size of the antenna shown in Figure 4.1).

Figure 4.5 — Modeled impedance data for the ¹⁄₄₀ wave vertical antenna of Figure 4.4.

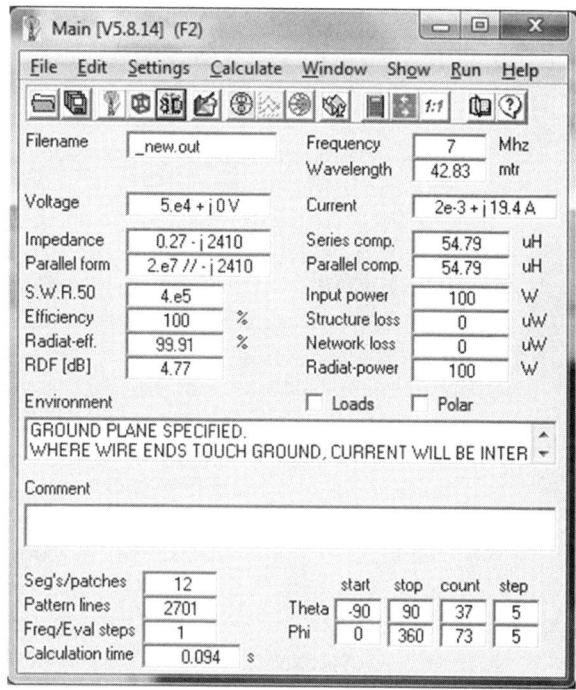

Figure 4.6 — 3D radiation pattern of the ¹⁄₄₀ wave vertical of Figure 4.4.

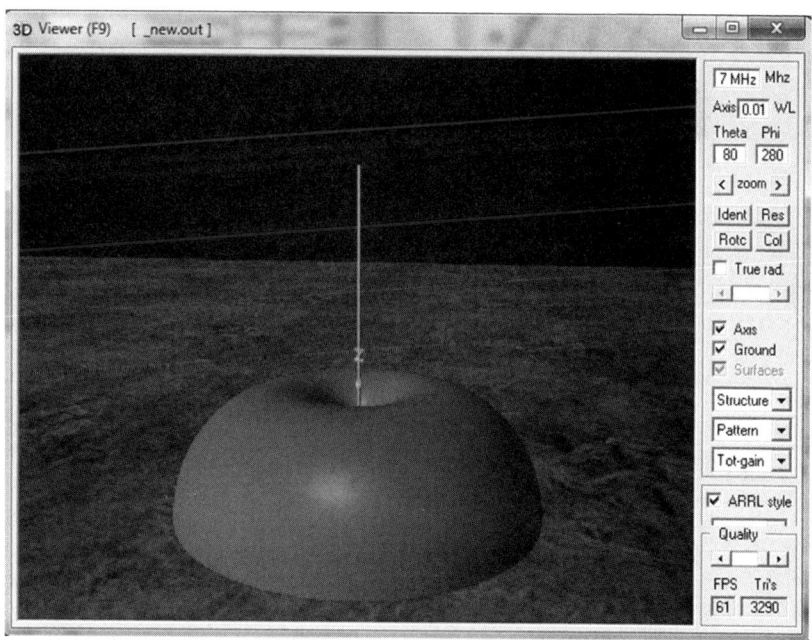

while the capacitive reactance is huge. In fact, for all practical purposes, the antenna is nothing but a very tiny capacitor!

Interestingly enough, as you can see in Figure 4.6, the radiation pattern has not changed one iota…at least in shape. The *size* (gain) is a lot smaller, but the radiation pattern is identical. (It's geometrically similar but not congruent.)

So, just how much smaller *is* the gain? This is the figure given in the RDF field of the output. For the "full size" antenna the RDF value Figure 4.2 is 5.17 dB. For the "shrunken" antenna, the RDF value Figure 4.5 is 4.77 dB. A difference of a measly 0.4 dB!

Now are you surprised? You should be ecstatic! A 1/40 wave antenna has less than a dB of "loss" over a 1/4 wave antenna.

Now, if something seems wonky here, we *did* leave out one small factor: the resistive loss of the wire. We have not accounted for the loss of *efficiency* for a very short antenna. In such a case, the "ohmic" loss is a much greater proportion of the total resistance, which is why antennas with higher radiation resistance are more efficient, all other things being equal. But the loss of *gain* incurred by shrinking our antenna, what we might call *geometric gain*, is nearly inconsequential.

The good news is that *efficiency* for a receiving antenna is far less important than efficiency for a transmitting antenna. So as we explore the "shrunken antenna" further, please keep this lesson in mind. We'll refer to it almost continually.

Now, just in case you still aren't convinced about the "shrinkability" of a whip, we've gone ahead and modeled the famous HTTI (3-inch screwdriver antenna shown in Chapter 3) at 7 MHz (**Figure 4.7** and **Figure 4.8**). As you can see in the impedance readout, the radiation resistance is almost nothing, about 1/1000 of an ohm, and the capacitive reactance is about 20,000 Ω. The radiation resistance has decreased dramatically and the reactance has increased dramatically, as expected. But notice the gain is unchanged from the previous model, about 4.77 dBi. (It's probably a little lower a couple of decimal places out, but for all

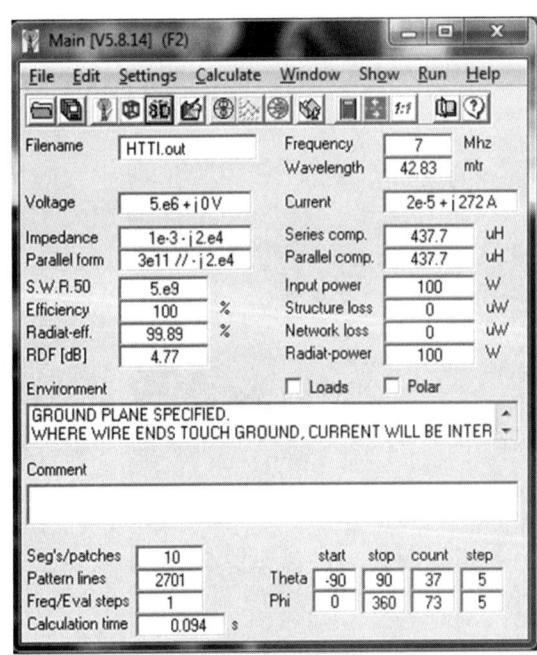

Figure 4.7 — Modeled impedance data for the 3-inch Highly Technical Test Instrument (HTTI) as an antenna at 7 MHz.

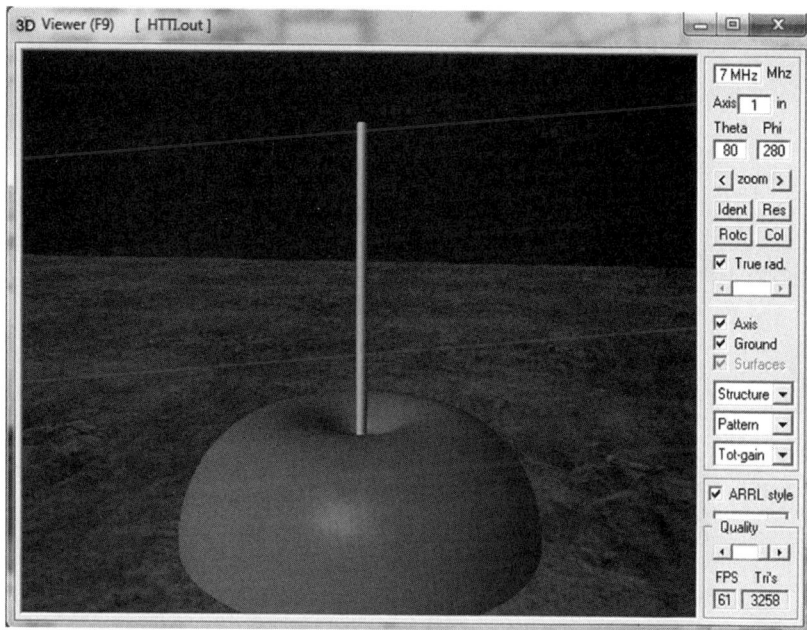

Figure 4.8 — 3D radiation pattern of the 3-inch Highly Technical Test Instrument (HTTI) of Figure 4.7.

practical purposes, it's unchanged.) Also notice the radiation pattern is unchanged.

In our next chapter, we will describe the "wave interception mechanics" of a "normal" receiving antenna, and then follow that up with the unique characteristics of a very short whip.

Chapter 5: The Receiving Antenna as a Signal Generator

Babbling Brooks and Spinning Saws

To properly think about the antenna in its receiving role, it is helpful to begin with a somewhat folksy analogy. Imagine an American Colonial era sawmill, powered by a waterwheel in a nearby babbling brook. A shaft comes from the waterwheel in the babbling brook, in through the wall of the shed, which then turns a saw blade, either directly or through a belt or two. Now, if you're a worker inside the shed, you may look at the shaft coming through the wall, and come to the conclusion that the shaft is the source of power. You may or may not have a clue what's on the other side of the wall; all you know is that the turning shaft supplies power to your saw blade.

However, those of us on the outside of the shed can clearly see that there's a waterwheel partially submerged in a babbling brook. We know that it is actually the babbling brook that's supplying the power, not the shaft.

We also might safely conclude that the waterwheel is taking only a small sampling of the total power available from the babbling brook. Indeed the vast majority of the babbling brook goes around the waterwheel. Unless it's a very exceptional waterwheel, someone a few thousand feet downstream along the babbling brook will probably not know one way or another whether your waterwheel is doing anything or not. It has an inconsequential effect on the babbling brook, though to the worker inside the shed, it's a rather crucial element.

Now, from a naïve point of view, the saw operator would reasonably consider that the twirling shaft is the source of power. After all, it is what makes his saw spin and do real work like cutting wood. He would probably also recognize that there's a finite amount of power available from the spinning shaft; it may noticeably slow down as he works through a large or stubborn plank.

It shouldn't be too difficult to substitute the twirling water wheel shaft with our receiving antenna. As far as our receiver is concerned, the antenna is a *source* of power, though we should recognize that it's not the *primary* source power; that is actually a transmitter tens, or hundreds, or even thousands of miles away. That remote transmitter is utterly unaware of the existence of our receiving antenna. Now, although, as we mentioned in Chapter 2, the amount of power our receiver actually absorbs from our antenna is truly minuscule, it *is* finite. We can actually *measure* the work that that signal does…in effect, our receiving antenna is now a signal *generator.*

Maximum Power Transfer Theorem Revisited

Being an electrical generator, the receiving antenna is subject to the same constraints as any other electrical generator. We know from basic ac electronics that the maximum power is transferred to a load when the load impedance is equal to the internal impedance of the generator. (For the sake of this discussion, we will consider the source and load impedances as being pure resistances. The slightly more complex *conjugate match* theorem, which includes reactive sources and loads, will be discussed later.)

Translating the maximum power transfer theorem to our antenna, we will convey the most power possible from our antenna to our receiver when the input impedance of the receiver is equal to the source impedance of our antenna. But what exactly *is* the source impedance of our antenna? It is that mysterious entity, *radiation resistance.* In free space, the radiation resistance of a dipole is 72 Ω. At typical heights above ground the average HF dipole combined with its image in the ground, has a radiation resistance of 50 Ω, a good match to our 50 Ω receiver input.

Now, the interesting thing about the maximum power transfer theorem is that it does *not* imply that the matched state is the most efficient state. In fact, any perfectly matched generator and load will be only 50% efficient, as exactly half the power will be dissipated in the internal resistance of the generator. (For the sake of this argument, we can temporarily disregard switched mode power supplies or even such devices as Class C RF amplifiers.)

How do we reconcile this with our receiving antenna? Assuming there is no "ohmic" resistance, the only source resistance we have is radiation resistance. But what actually happens when we have a perfectly matched antenna and receiver? As in the case of the "normal" electric generator, half the power will be consumed in the receiver, and the other half will be *re-radiated.* This should not be too surprising; it is completely compatible with the maximum power transfer theorem.

It needs to be reiterated that the gain (or directivity) of an antenna and the efficiency can be two very different things. One common example of this is in the case of a directional AM broadcast station, normally consisting of a phased array of two or more verticals. The FCC broadcast rules assume that each additional radiator will decrease the overall efficiency by 5%, and so allows an additional 5% of array *input* power for each additional radiator to compensate for the inevitable losses. In most well-designed transmitting and receiving antennas, however, the loss of efficiency is an acceptable tradeoff for increased gain…or some other useful property.

Well, back to our matched receiver for a moment. In most typical situations, our goal is to transfer as much of the minuscule intercepted power to the receiver front end as possible. This occurs when the radiation resistance is equal to the receiver input resistance, as we've already determined. This will result in the greatest S-meter reading for a given transmitter radiated power (considering the propagation path in between transmitter and receiver). But this condition *also* results in the greatest transfer of *noise* power. This is easy to demonstrate if you use an external antenna tuner with your transceiver. You can definitely hear the noise increase as you adjust the tuner for the proper impedance match (and resonance). You have probably also noticed that if you adjust your transmatch for maximum noise level, it usually results in a *pretty good* setting for minimum transmitter SWR as well. What this tells us is that your receiver input is indeed *pretty close* to 50 Ω, as we will discuss in some detail later.

Modern solid-state receivers have much more uniform input impedances across the HF spectrum than most "boat anchors" of times past, not that this results in any significant performance advantage. It simply means that, when using the same antenna for transmitting and receiving, there's less adjustment needed. There is nothing sacred about 50 Ω as far as receiver performance is concerned. Indeed, many old TV receivers had 300 Ω inputs — a good match to low-loss parallel line and folded dipoles — but now most sport 75 Ω.

Striking a Resonant Chord

It should be noted here that there is nothing particularly sacred about a *resonant* antenna, either. The ability of an antenna to radiate (or extract) power from the ether has nothing to do with whether the antenna is resonant or not. Where resonance *does* come into play is when we are attempting to match said antenna to a load. Power is only consumed (or generated) by the *real* component (resistive component) of the circuit.

This is an important fact to keep in mind as we develop the highly effective *aperiodic* active antenna. In an aperiodic antenna, no effort is made to resonate the antenna or its associated amplifiers. The Beverage, Waller Flag, and other popular designs are non-resonant receiving antennas.

Conjugal Rights

At this time, we need to take a brief but important side trip from the matter at hand, determining the actual input impedance of a typical HF receiver. This will necessarily take us into the matter of the *conjugate match,* a concept well covered by authors such as the late Walt Maxwell, W2DU, but well worth revisiting.

For most practical operating purposes, precisely knowing (and matching) the input impedance of a modern HF receiver is not too important, as most receivers of late vintage have more than enough gain even in a severely mismatched state. (We won't paint this assertion with too wide a brush, as at VHF and UHF, receiver impedance matching becomes much more crucial.) But for the most part, fairly haphazard impedance matching on the receiving end of an HF station is not generally a show-stopper. However, when it comes to actually *measuring* the performance of a receiver, antenna, or both, knowing the actual impedances we're working with is of prime importance. If we wish to accurately ascertain the gain (or loss) of an active antenna — or any antenna, for that matter — we will need to know the impedance at which we're making voltage measurements. While, as suggested earlier, the input impedance of a modern HF rig is *reasonably* close to 50 Ω across its operating range, it's nice to be able to (or at least know how to) determine this impedance.

But even better, we can create a new, precise, 50 Ω front end for our receiver by means of an antenna tuner (a matching network). (We do want to be sure the matching network has minimal losses.) Although the very point of an active antenna is usually to *avoid* extra tuned components like antenna tuners, such devices are invaluable when making actual measurements. When an antenna tuner is inserted in front of a receiver, the input side actually *becomes* the "front end" of the receiver. The conjugate match transfers the existing input impedance of the receiver to exactly 50 Ω, non-reactive. We can then confidently measure the voltage at the 50 Ω input of the antenna tuner and thus derive accurate relative or absolute power measurements.

A Nifty Procedure

While you *can* measure the input impedance of a receiver using a vector network analyzer (VNA), or even a more common antenna analyzer, these instruments can put out significant power, which your receiver's sensitive front end may not truly appreciate. Instead, we will use an indirect method, commonly known as the *substitution method,* using an antenna tuner, whatever existing receiving antenna you may have, and a signal generator of some kind. I've found that an antenna analyzer combined with my exclusive Highly Technical Test Instrument screwdriver antenna serves as just the right weak signal source…enough to generate a robust signal without frying the receiver's front end. (See **Figure 5.1**.) We will then optimize the antenna tuner for the maximum received signal strength.

For the actual measurement you *can* use your S-meter, but it's preferable to disable your AGC (the AGC voltage is what actually drives the S-meter), and then measure the audio output signal with an oscilloscope or an ac voltmeter. Most modern digital multimeters (DMMs) measure ac voltage accurately into the upper audio frequency range.

First choose a middle range amateur frequency, perhaps 10.120 MHz. (I am always careful to be within an actual amateur band even when using a wimpy signal generator. Although I've never known anyone to be given an FCC pink slip for a dip oscillator or antenna analyzer operating out of band, it's theoretically possible. Any time you radiate *any* radio signal, you must be duly licensed…with a few rare exceptions.)

Connect the output of a high quality antenna tuner to your receiver's antenna input with a short length of 50 Ω coax. (The actual direction you connect your tuner normally doesn't matter, as long as you're consistent. However, if you're easily confused, always connect your tuner so the signal *source* is at the tuner's "official" input terminal.) Connect your receiving antenna and normal operating coaxial feed line to the antenna tuner's input. Just to avoid any possible confusion, we'll number the ports 1 and 2.

Position the antenna analyzer with the HTTI affixed beneath your receiving antenna. Set the frequency to 10.120 MHz. Return to your radio room, turn on your receiver in CW mode, and dial in the test signal to obtain a comfortable tone in your speaker. Again, if you can disable your S-meter and measure the audio level, that is best, but if you can't you can use your S-meter. If your S-meter is pegged, of course you will need to move the antenna analyzer farther away from your receiver antenna — or use a shorter

Figure 5.1 — An antenna analyzer with screwdriver antenna attached provides enough signal for receiver measurements without damaging the receiver front end.

The Receiving Antenna as a Signal Generator

HTTI antenna. Any readable signal below S-9 will work. Do not readjust anything on the receiver after this point.

Now, carefully adjust the antenna tuner for maximum received signal strength, using the S-meter or an audio meter. When you are satisfied that you have the "sweet spot," go outside and retrieve your antenna analyzer; you'll need it for the next measurement! Do not readjust the frequency on your antenna analyzer.

Being very careful not to disturb the antenna tuner settings, disconnect the coax connector from your receiver's input terminal, and connect it to the output of your antenna analyzer. Carefully read and record the resistance and reactance that your antenna analyzer displays. This value is the *complex conjugate* of your receiver's input. If your receiver's input is indeed 50 Ω, non-reactive, your antenna analyzer's readings will be the same. Most likely the reading will depart slightly from the ideal 50 + j0 Ω.

So, what is the complex conjugate, anyway? This is quite simple. The complex conjugate of an impedance has the same value as the resistive component and the same value but opposite sign of the reactive component. For example, if your antenna analyzer gave you a reading of 60 –j30 Ω, the complex conjugate would be simply 60 + j30 Ω. Now, if you've been paying attention, you would recognize that complex conjugate matches are always resonant, since the reactances are equal and opposite. Are we impressed yet?

Even Trickier

It's always a good idea to do as many "sanity checks" as you can on any measurement method. Here's a little trick that may be a little surprising, but really shouldn't be. It does require one more ingredient: a good non-inductive 50 Ω resistor (a good dummy load). With your antenna analyzer still connected to your antenna, readjust the antenna tuner to get precisely 50 Ω, 0 reactance. Now, remove your antenna analyzer from the antenna tuner, and connect your 50 Ω load resistor in its place. Move your antenna analyzer to the other side of the antenna tuner, and again, without disturbing the tuner adjustments, take an impedance reading. The value you see here is the complex conjugate of your *antenna* impedance. (Actually the impedance of your antenna and transmission line combination *after* that point.) How do you confirm this? Very simple. Make a measurement of your antenna feed line directly! It will be the complex conjugate of the antenna tuner with the load resistor on the other end.

Consistency

At this point you probably know more than you needed to know, but this information will become very handy later on. The substitution method is adaptable to many different kinds of measurements, and it has the added feature of making some kinds of errors cancel out.

Power Factor

If there is a good reason to obsess about antenna resonance, it has very little to do with radiation efficiency, but it has everything to do with ease of calculation. The feed point impedance of an antenna (or actually an antenna *system*) that is resonant has a *power factor* equal to 1. Now, for some strange reason, you don't hear the term power factor used much by radio folks, whereas in the electrical power industry, it's a well-understood concept.

There's nothing fundamentally different between radio and electrical power transmission…other than that in the electrical power realm you generally try to *avoid* radiating power into space. But the concepts in a confined *circuit* are identical. In a circuit with a power factor of 1, the true power and apparent power are the same. All our reactances cancel out, and we can treat every component as a resistor. If the input impedance of a receiver is *resonant*, that is, if it has a power factor of 1, we can easily calculate the true power into the receiver using E^2/R. It's a little bit messier if we have to figure in reactances.

The Resonance Fallacy

Many hams are surprised and appalled to learn that most AM broadcast towers are *not* self-resonant…in fact most are far from it! AM broadcast tower sections typically come in 20 foot, or sometimes 60 foot, sections. If you're *lucky*, you can build an AM tower that's within 20 feet of resonance. *Nobody* cuts AM broadcast towers to the "proper" length… nor have they ever! This means that the typical AM broadcast tower is quite reactive. It is up to the antenna tuning unit located in the adjacent "dog house" to make up the difference.

The Active Antenna as a Voltage Probe

We will take the mystery out of the surprising performance of the active whip antenna by treating it as a simple voltage probe. We will demonstrate how a true *voltage follower,* the heart of an active antenna, is capable of *infinite power gain.* We will also demonstrate how the low thermal agitation noise of a very short whip can result in a *better* overall signal-to-noise ratio than a full sized HF antenna.

The operator of every AM station measures power by the I²R rule. This means he or she has to know the radiation resistance of the tower (but not the reactance!) and the RF current going into that radiation resistance. An RF ammeter is inserted *after* the tuning circuitry, so it will always see the true radiation resistance, not a converted or "matched" resistance. Although it is true that the maximum current will flow when the antenna *system* is resonant, as far as calculating actual radiated power, the reactance is irrelevant! I²R with an RF current meter always gives you *true* power.

This principle is one of the antenna laws that is truly reciprocal in nature. When a receiving antenna is connected to a load (the receiver), it is *only* the resistive part of the load that consumes any power. Reactance will increase the net *impedance*, making it more difficult to have the desired current flowing in the antenna resistance, but the true consumed power is only determined by the resistance.

As far as the receiving end goes, this is actually very good news, because if we *aren't* concerned with power factor, we can arbitrarily shrink our antenna to any size we like. In fact, the *radiation resistance* of an antenna continually decreases as it becomes smaller, while the series reactance (dipole) continues to increase or the parallel reactance continues to decrease (loop). A very short whip antenna on the lower HF bands has essentially zero power factor! But you might fairly ask how this could possibly be good news.

Power factor is only important if you need some power. But an active whip antenna is essentially a *voltage* sampler. It simply takes a *sample* of the E-field voltage. It simply needs to "know" what the voltage is at any point in time; it doesn't have to actually *do* anything with this information. It is the job of the active amplifier (preamplifier) to do that. The active amplifier converts a voltage to a power, sometimes at a considerable lower voltage than the original sample — but with much greater current — sometimes approaching *infinitely* more current.

to a high impedance antenna, allowing the antenna to develop the same voltage as a wire in free space....or very nearly so. Beginning with the ready availability of FETs in the late 1960s, designs of a number of active antennas flourished, both of the tuned and the untuned variety.

Taking a look back through the archives, we find a typical active antenna consisted of a single FET, in source follower mode, followed by a high gain, conventional transistor (BJT) amplifier of three or four stages (typically with a lot of negative feedback for stability and bandwidth). Total preamp gains on the order of 50 dB were used, which just about compensated for the normal loss of a very short antenna (compared to a dipole). This is still a good benchmark to shoot for.

Incidentally, if you really like twiddling numbers, there is an excellent presentation on some of the math involved in short antennas by the Dr. Ulrich Rohde, N1UL. See **synergymwave.com/articles/2016/Antenna_presentation.pdf**.

Follow This

The common drain FET amplifier, also known as the *source follower*, is a direct descendant of the vacuum tube *cathode follower*. In the ideal source follower, the source current (as appearing across a source load resistor) precisely follows the input voltage applied to the gate. In reality, the output voltage will always be slightly less than the input voltage). If we make the assumption that the output voltage *is* the same as the input voltage, then we know that the power gain will be equal to the *input impedance* divided by the *output impedance*. If the input impedance is infinity, the power gain will be infinite if the output impedance is *any* finite value. Infinity divided by anything other than infinity is still infinity!

In reality, an FET gate's impedance will never actually be infinite, but it can be very, very high, up to a thousand gigaohms in the most modern devices. Even if we include incidental circuit component losses, it is very easy to build a source follower with an impedance of a million ohms and an output impedance of a thousand ohms. This gives us a 1000:1 (or 30 dB) power gain in just one stage. This far more than compensates for the loss of gain incurred by our self-imposed "antenna shrinkage," even on the lowest HF bands.

New Devices

While the benefits of the FET source follower have been known and employed for many years, there has been a perception among many hams that preamplifiers of any kind should yield some kind of *voltage gain* to be meaningful. It needs to be re-emphasized that it is the *power* that does

Some Relevant Op-Amp Theory

Only in recent years has the ubiquitous operational amplifier, o\
truly suitable for radio frequency applications. However, as amazing
new devices are, you can't haphazardly drop them onto a circuit boa\
high performance active antenna to magically come to life. There are s
tricks and disciplines necessary to making RF op-amps work properly. W
you around all (or at least *most*) of the potholes in this new territory. Specia\
be given to the Analog Devices AD8067 op-amp in this overview.

the real work in any radio circuit (or any physical device, for that matter). Voltage is never an end to itself; no matter how much voltage gain we might achieve in a receiver chain, until we convert that voltage into actual power, it's not going to do us much good. The lesson here is, don't be too disappointed of you don't see a large voltage gain between the input and output of your painstakingly crafted preamplifier. Don't be too impressed if you see a lot of voltage gain, either. Remember, the bottom line is always *power gain*.

While there has always been a somewhat limited variety of FETs available for weak signal radio applications, the availability of decent radio frequency op-amps has been practically nonexistent, until quite recently. For the most part, op-amps, the building blocks of most modern analog circuitry have been mostly limited to low frequency and audio instrumentation. Although versatile RF amplifier chips like the CA-3028 have been around for a while, they fail to qualify as true op amps on a number of credentials.

A number of late arrivals have changed this situation, however, much to our joy. As an example, the Analog Devices AD8067 FET input op amp is a *true* radio frequency op-amp, with an impressive gain bandwidth product (GBP) of about 350 MHz. This device will operate "flat" to 50 MHz at a voltage gain of 10. Combined with the extremely high input impedance of the FET input stage, this little gem can essentially function as "a voltage follower with gain" if I might coin an original definition. It is supremely suitable to any active antenna, and is the "star" of several projects described later in this book.

A Few Cautions

We will save you some heartburn by revealing some lessons we learned the hard way, when using the remarkable AD8067. Actually, the practices we describe apply to *any* radio frequency layout, but things you might have gotten away with before, you *can't* get by with when using a

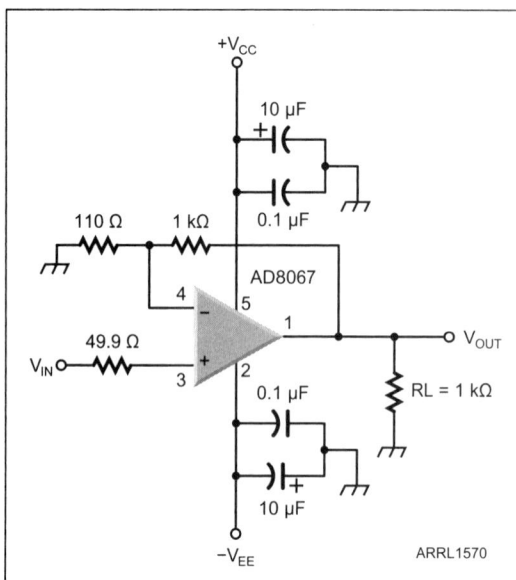

Figure 6.1 — A typical non-inverting 10:1 gain amplifier using the AD8067.

chip like the AD8067. Its incredibly high gain bandwidth product will almost guarantee that you experience instability of some sort if you try to take construction shortcuts. Here are some rules you *must* follow:

1) Use an effective ground plane. This precludes any kind of breadboarding or perfboard construction. Use double clad copper board, removing as little copper as possible between the chip terminals. Bond the upper and lower cladding of the board together electrically.

2) Bypass the input voltage terminals using capacitors with very short leads. A combination of 10 µF and 0.01 µF capacitors on each lead seems to be universally effective.

3) Separate the non-inverting input (pin 3) as much as possible from the output lead (pin 1), preferably by leaving a small ground trace to remain between them. Avoid any path for capacitive feedback from pin 3 to pin 1.

4) A 500 kΩ to 1 MΩ resistor needs to be applied between pin 3 and ground, *unless* you want to use *guarding techniques*. We have not seen the need for guarding methods for any of the projects in this book as long as the aforementioned resistor is installed

Figure 6.1 shows a typical non-inverting 10:1 gain amplifier using the AD8067. We have found that this particular configuration is quite forgiving and stable when laid out according to the rules described above. It is the configuration used in most of the active antenna projects described in this book. "If it ain't broke, don't fix it!"

There are other good layout practices to follow in any RF design work, but these few rules should keep you out of trouble with our "star" AD8067. We will describe RF layout methods in more detail in the actual construction projects.

The bottom line is, even a silver bullet (which we consider the AD8067 to be, when it comes to active antennas) needs to be fired from a trained hand!

Chapter 7

The Role of the Resistor in the Receiving Antenna

As a general rule, resistors are things you want to avoid in any transmitting antenna or system. Resistors consume precious RF power, converting it into heat, which is something we'd rather not do. Various schemes exist for "broadening" the bandwidth of transmitting antennas using resistive loading of some kind. While this may make your transmitter "happy" over a wider variety of conditions there are better ways of achieving wide bandwidth without "burning up" power. Even the venerable rhombic and V beam antennas, which traditionally use termination resistances, can benefit by using more efficient "recirculation" methods. But that will not be the topic of this chapter. Here we will talk about the use of resistance in receiving antennas to achieve things that cannot be achieved any other way.

When is an Antenna Not an Antenna?

The simple answer is: when it's a transmission line.

As it turns out, nearly every imaginable antenna also has some transmission line properties, at least to some degree. In many cases, we can ignore the transmission line characteristics of an antenna, in some cases we cannot.

Resistive Loading of Receiving Antennas

As just about any experienced ham can tell you, the use of *resistive* loading elements of any kind in a transmitting antenna is something that should be avoided like the plague. Adding a dummy load to an antenna is *not* a suitable way of increasing transmitting bandwidth. On the other hand, resistive *loading* and *isolating* are extremely useful techniques in numerous *receiving* situations, both for increasing usable bandwidth and manipulating the patterns. Since *efficiency* is generally *not* a priority in receiving antennas, we are free to use resistors in our antennas to achieve a number of special performance characteristics.

Let's first look at the extreme case of a folded dipole. In most incarnations, the folded dipole is actually *made* of transmission line, such as 300 Ω twin lead. The impedance transformation property of the folded dipole (that is, its ability to raise the input impedance to 300 Ω from its nominal 75 Ω) is a property of the transmission line "stubs" of which it is made. It is important to note that this feed point resistance is *not* the same as the radiation resistance, when using such a transformer. Radiation resistance is fairly rigidly tied to the physical size (aperture) of the antenna. Tom Rauch, W8JI, gives a very concise dissertation on this topic on his website (**w8ji.com/radiation_resistance.htm**). However, it also radiates just like any dipole...very much unlike a balanced transmission line. In reality, the full analysis of the folded dipole is fairly complex, and one must look closely at both the relative currents *and* their phase angles to understand how it can actually radiate.

A less extreme example is the lowly end-fed long wire. As the antenna approaches a number of wavelengths, a few things happen. The radiated power from the antenna becomes slightly less as you move away from the source. Although some of this can be attributed to copper loss in the wire itself, it is primarily due to the fact that energy is being radiated along its entire length, and there is, logically enough, somewhat energy left to radiate when it finally gets to the end.

While the gradual loss of energy from the source to the free end of the wire distorts the pattern slightly, the antenna is, for all practical purposes, bidirectional. It will have two primary lobes, which are conical structures surrounding the wire, with an angle between the wire and the lobe being increasingly less as the wire becomes longer.

However, the wire also behaves like a transmission line. How *much* like a transmission line it behaves is a function of its length, in terms of wavelength. But any antenna that is long enough to have gain *along* the length of the wire, or close to it, will also be long enough to exhibit transmission line properties. In most cases, the other half of the transmission line is the Earth; a long wire at moderate height will have a characteristic impedance of between 100 and 600 Ω, measured between it and the ground.

If this transmission line is terminated in its characteristic impedance, such as with a 600 Ω load resistor, it will act as any other transmission line. It will be *flat* — in other words, the standing waves along the line will disappear (and the line becomes a "traveling wave" antenna). It will therefore become a non-resonant or *aperiodic* antenna, either for transmitting or receiving.

But another interesting thing happens. Transmission lines can behave as *directional couplers* — they can be made to induce currents only in the direction that a radio wave is traveling. This is probably the more important property of terminated long wires and their cousins, as far as receiving is concerned. With the proper choice of termination resistor, the terminated long wire can achieve an impressive front-to-back ratio, which can also be switched simply by swapping the termination and feed-point connections.

It should be re-emphasized that any resistor in any antenna system will result in *real* power loss, either in transmitting or receiving. However, in receiving applications, the loss of power due to resistive loading will be generally offset by a much greater gain in some other performance area. For example, as in the case above, the resistor can "burn up" power arriving from an undesired direction.

Going Through a Phase

Another worthy application of resistance is in the control of *phasing* of an element in a receiving array. In some instances, such as direction finding or interference nulling, it is necessary to carefully adjust both the phasing and relative power levels of two or more antenna elements independently. This can often be difficult or impossible to do with just variable reactances. Various *bridge* circuits using resistance and reactance can be incorporated in a receiving antenna to achieve nulls in some desired direction.

A somewhat curious antenna we will discuss later is the *aperiodic loop*, a compact antenna that is often combined into large arrays. An aperiodic antenna is any antenna where resonance effects are relatively unimportant. In most cases, an aperiodic antenna is very small compared to a wavelength, and is often resistively loaded to further reduce any standing waves. The very short whip described earlier may be considered an aperiodic antenna, but for the present discussion, we will concentrate on resistively loaded, short antennas, most commonly in a closed loop form. Unlike most small loops, which are tuned, the aperiodic loop is a broadband antenna. In addition to giving the antenna a wide bandwidth, the resistive termination of an aperiodic loop can be used to closely control or modify the radiation pattern. Aperiodic loops are normally operated as *active* antennas, since, because of their small size, they have considerable loss.

Winning the Pennant

A number of closely related antennas, such as the pennant, the Waller Flag, the EWE, and the popular K9AY Loop incorporate resistive loading. All of these antennas are known for their low noise performance, especially on the lower HF and MF bands, and also have the capability of producing deep nulls in some direction. These antennas all operate in the "gray zone" between antenna and transmission line, where the load resistance is an integral part of their makeup.

While the EWE is largely ground dependent, as is the Beverage, the Waller Flag, the pennant, and a number of other "flag-like" antennas are known for being relatively ground independent. See the July 2000 *QST* article, "Flags, Pennants, and Other Ground-Independent Low-Band Receiving Antennas" for a good overview. The Waller Flag deserves a special mention because it is a bit more complicated than other flag antennas, but is worth the effort to construct when faced with challenging noise problems.

A Model Citizen

While we will, later on, devote a large section to receiving antenna modeling, we need to say a few words in particular about antenna modeling with resistive elements. I am particularly fond of the *4nec2* antenna modeling program (**www.qsl.net/4nec2/**), and not just because it's entirely free…though that is a consideration. *4nec2* is based on the original Lawrence Livermore Numerical Electromagnetic Code *NEC2* engine. The extremely popular *EZNEC* is based on the *MININEC3* engine.

There are advantages to each type of engine. *MININEC3* has some advantages in modeling real world ground systems, for example, while the *NEC2* engine does a better job of modeling things such as load resistances and non-uniform conductor diameters. In particular, the optimization capability of *4nec2* makes it a lot easier to optimize resistive elements for given performance criteria. However, all flavors of *NEC* work on the same Method of Moments algorithm, and you should obtain very similar results regardless of which variety you use. The important thing is that, no matter which modeling program you use, you should subject your model to some kind of "sanity check," especially if you are working with a very unusual design.

There is a nice selection of example receiving antennas included in the *4nec2* package, under the directory Aperiodic. These models can show you just how sharp a null can be in a well-designed flag, pennant, or EWE antenna. We highly recommend just playing around with these "prefabricated" models to get a feel of what termination resistors do for the pattern.

As with *any* modeling program, the best model is never a valid substitute for actual thought. You should have a fairly good idea of what an antenna should do before turning it over to your computer. Good antenna models seldom yield any real surprises; where they are most useful is in doing fine tuning and optimizing. Any antenna model that gives you an answer that seems too good to be true almost certainly is.

If you haven't done any antenna modeling before, perform a reality check of a simple antenna like a dipole, where you *know* what the pattern should look like.

Noise That Annoys

One thing you need to be aware of is that resistors of any kind always add *thermal noise* to a system. It would be sort of dumb to add resistance to an antenna, to supposedly increase the noise immunity in some direction, if in so doing you create more incidental noise than you cured. Resistors are always a bit of a Catch 22 in this regard. At what point does the addition of a resistor in an antenna contribute significant noise?

We can't give you a definitive answer right here, but we can point out some trends. We can imagine any resistor (at some given temperature) as being a source of noise power, so it generally follows that whatever condition results in the best match between our resistive noise source and our load is the thing we want the *least*. The maximum power transfer theorem works for *noise* as well as "wanted" signal. So it should be evident that we have somewhat of a balancing act to perform. We want the best transfer of power of desired signal consistent with the lowest level of *noise power transfer*. Or, put another way, we want an impedance *mismatch* between any noise generator and our load (presumably our receiver's front end) *and* a reasonable impedance match for our desired signal. In general, low values of series resistance contribute less overall noise to an antenna system. Where you must use resistors in a receiving antenna, (or even associated circuitry, such as in a preamplifier) always use the appropriate amount of resistance that will allow the antenna to function consistent with acceptable noise performance.

There are a few tricks to reduce resistive generated noise (otherwise known as "thermal noise" or Johnson-Nyquist noise — as distinct from receiver atmospheric noise), which we will discuss later on.

Chapter 8

The Small Loop Antenna

The small loop antenna, while being a somewhat specialized antenna, merits an entire chapter because of its importance in receiving applications, especially at the lower Amateur Radio frequencies. A conventional half-wave dipole at resonance supports a classical standing wave. The current distribution is a sine wave with no current at the ends, and maximum current in the middle. We can also look at the voltage distribution, which is also a standing sine wave, but with peak voltages at the ends and a minimum voltage at the center.

The current distribution along a dipole results because the *impedance* at any point along the wire is different. It has essentially infinite impedance at the ends, and approximately 72 Ω of feed point impedance in the middle. If we constrain the antenna to operation precisely at resonance, we can ignore *reactive* components. We can look at the dipole as having a variable impedance along its length. It's important to stress that voltage is a *difference of* potential between two points, and in many cases there may not be an obvious "second terminal" to measure voltage against. In the case of a dipole, the voltage can be the difference of potential between two points on either side of the center. Or in the case of a ground plane (Marconi), this can be the voltage relative to ground. In either case, it's the voltage *distribution* that is most telling, as well as the most familiar. This familiar voltage distribution only occurs when the wire is a significant fraction of a wavelength, in this case ½ wave.

In the case of a "magnetic" antenna (**Figure 8.1**), the current around a closed

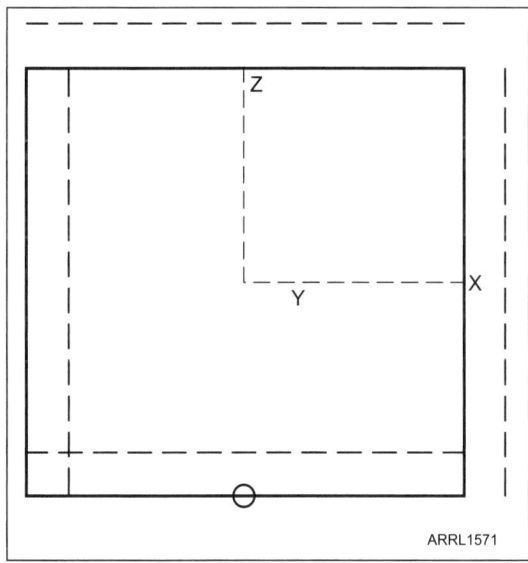

Figure 8.1 — In the case of a "magnetic" antenna, the current around a closed loop is considered to be basically uniform. Here is a loop antenna that's a mere 0.01 wavelength on a side. The dashed lines display the current magnitude.

The Small Loop Antenna **8.1**

The Loose Coupler

As we have already demonstrated, in an active antenna we make no effort to achieve an impedance match between the antenna and the following circuitry. In a typical active antenna, we can make up for antenna "deficiencies" with a high quality amplifier in order to achieve an improvement in bandwidth, convenience, or signal-to-noise ratio.

However, the usefulness of "mismatched" antennas is not limited to active antennas. A tuned loop antenna is often very *loosely coupled* to the following circuitry (namely a receiver) in order to maintain the high Q of the loop. This method and its associated hardware, known as the *loose coupler* (for naturally obvious reasons), was used since the earliest days of radio, when the passive crystal detector was king. Loose coupling means that the antenna system is *not* designed for maximum power transfer; indeed only a small fraction of the power that could be extracted from the antenna is actually extracted. Of course, this method does reduce the S-meter reading, but most receivers have more than adequate gain to make up for the mismatch. What one *does* gain with a loose coupler is very high selectivity and optimum dynamic range…which can translate to a greatly improved signal-to-noise ratio…our bottom line of antenna "goodness."

---

loop is considered to be basically uniform. In this example, we have a loop antenna that's a mere 0.01 wavelength on a side. The dashed lines display the current magnitude.

The use of the term "magnetic antenna is a bit misleading, however, as *every* antenna has both a magnetic and electric field. However, the popular concept of a "magnetic loop" antenna is one where the circulating current is very high and nearly uniform everywhere, which is the result of the antenna being very short relative to a wavelength. Large values of capacitance are used in these antennas to bring them to resonance.

Going Through a Phase

For the time being, let us disregard the efficiency (or lack thereof) for a very small loop antenna, and focus on the directional characteristics. Full-sized loops, generally speaking, serve a rather different purpose than very small ones, and have quite different directional characteristics. For all practical purposes, a full wave loop can be compared to a pair of stacked dipoles. Again, the current distribution is shown in dashed lines in **Figure 8.2**. The major portions of the currents are in the horizontal components, and in phase. The vertical components contain much less current and are out of phase.

The crest of an incoming vertically polarized wave, arriving from the left, will strike the first vertical part of the loop (segment 1) slightly before striking the "downwind" (segment 2) of the loop. The slight

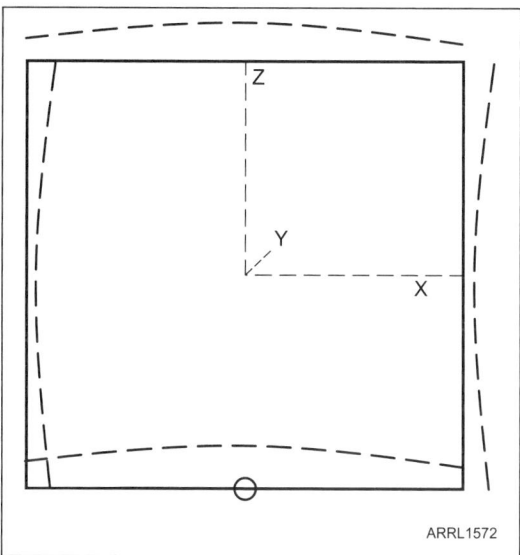

Figure 8.2 — For all practical purposes, a full wave loop can be compared to a pair of stacked dipoles. The current distribution is shown in dashed line. The major portions of the currents are in the horizontal components, and in phase. The vertical components contain much less current and are out of phase.

difference in timing between the arrival at segment 1 and segment 2 is minuscule…but existent. The larger the diameter of the loop, relative to wavelength, the greater will be the difference of potential of the two vertical segments. A wave arriving perpendicular to the plane of the loop will excite the two segments at precisely the same time, creating no difference of potential between them. There is, thus, no response to such waves, but rather a perfect null.

The signal you can extract from a tiny loop antenna is very small, and usually requires some kind of preamplifier. However, the perfect null allows for some good noise reduction in many cases, as well as providing precise direction finding capabilities.

Tiny loops can be more sensitive, with no change of their directional properties, by employing multiple turns under certain but not all circumstances. This is electrically equivalent to connecting many of the aforementioned sections in series, multiplying the difference in potential between the "upwind" and "downwind" wave portions. Even more sensitivity can be achieved by resonating the multiple-turn loop with an appropriate capacitor. However, this makes the antenna very narrow-banded, as well, which may or may not be what you want.

Some Polarizing Points

As we will discover, ground-mounted wave antennas tend to be vertically polarized, some very strongly so. This would not normally be true in free space, but as the height above ground is reduced, relative to wavelength, the effectiveness of many receiving antennas increasingly favor vertical polarization. (However, as soon as you elevate a Waller Flag or similar antenna more than a few tens of feet, horizontal polarization is preferred.)

True ground wave propagation, that is, propagation in which the wave is in contact with a conductive surface, *must* be vertically polarized. (By vertically, we mean normal to the surface of the conductor.) A horizontally polarized wave cannot propagate in contact with a conductor, as the electric field is shorted out. A vertically polarized wave, even if partially in contact with a conductive surface can propagate along the surface, however, but it will bend in the direction of the conductor. That is, vertically polarized ground waves will follow the contour of the ground. This is a rather handy phenomenon, as it allows ground wave radio signals to travel well beyond the horizon, with or without an ionosphere. However, the attenuation of ground waves increases with frequency. True ground wave propagation becomes relatively ineffective above about 5 MHz, and is thus, most suitable to the "low bands" on MF frequencies. Standard AM broadcasting relies heavily on vertically polarized ground wave propagation, at least during the daytime.

Now, it's been widely propagated in radio literature (pun intended) that most atmospheric electrical noise is vertically polarized, as well. This would seem to present a bit of a conundrum, to say the least. Decades of ham literature have suggested that horizontally polarized antennas are less susceptible to noise than vertically polarized ones. But is this really true? It certainly seems that most lightning induced static would be predominantly vertically polarized. After all, most lightning consists of ground-to-earth and earth-to-ground strokes. However, even if the main bolt of lightning is vertically polarized, and indeed does create a very strong dc or ELF vertical electrical component, this main bolt is not what creates most of the electrical noise at HF frequencies. The radio frequency static is primarily from the plasma "fuzz" branching off the sides of the bolt. Not to mention that there is a whole lot of cloud-to-cloud activity going on nearly continuously at some point on the planet.

Contrary to most modern "wisdom," manmade electrical noise is not predominantly vertically polarized...at least to start out with. However, since most sources of man-made electrical noise are both relatively low in frequency *and* emitted from relatively low angles, i.e. ground waves, the horizontal components tend to be shorted out, as described above.

The bottom line is that, at least statistically, low angle, low frequency *incoming* waves, whether noise or desired signals, will tend to be predominantly vertically polarized.

Therefore, using a horizontal antenna to reduce electrical noise is just as likely to reduce the desired signal as well. This does not hold as strongly, however, for higher frequency, higher angle-of-arrival signals, or in cases where you might have a receiving antenna high enough to be relatively immune to ground effects, which, of course, is only going to be practical on the higher HF frequencies.

For this reason, you will notice that most of the low noise receiving antennas, such as pennants, flags, EWEs and K9AY loops are all vertically polarized when used at ground level. This is no accident. However

Pennants, Flags, K9AYs, and Other Aperiodic Antennas

The extremely popular (and deservedly so) K9AY Loop is representative of an entire class of relatively small broadband antennas, which lend themselves nicely to *active* methods. Any of these antennas can also be incorporated into larger *arrays* of receiving antennas for outstanding performance.

there can be a significant advantage to using horizontal polarization once the antenna height (like a Waller flag) increases beyond 20 feet, for frequencies below 5 MHz.

The bottom line here is, don't fear or disparage vertical polarization at ground level! This may go against your "ham instincts" that tell you that *real HF antennas are horizontal.*

Making Sense of Sense Antennas

In a small loop, the current is essentially constant around the perimeter of the loop, unlike a full wave quad loop, which has a definite standing wave…and very different radiation properties. The small loop antenna is bidirectional. It receives signals equally for signals arriving from any edge of the vertically oriented loop (including straight up). You cannot identify the direction of the incoming signal by rotating the antenna for a maximum S-meter reading. Nor can you use the *null* at 90°, since there are two of these as well. The small loop antenna is *ambiguous.* How do we remove the ambiguity? We use a clever little device called a *sense* antenna.

The sense antenna is a small, omnidirectional whip antenna, located not far from the small loop antenna. A phasing and combining network allows you to mix the signal from the loop and the sense antenna in just the right phase to create a unidirectional *cardioid* pattern, with one sharp null, perpendicular to the loop. This will allow you to find the approximate direction of the remote station (or noise source), so you can eliminate anything that's 180° off. Then you can disable the sense antenna and readjust the loop for a perfect null. Most commercial noise cancelling antennas use this general method for eliminating noise arriving from a single point, and they can be extremely effective.

The small, tuned, shielded loop antenna has been the mainstay of RDF (radio direction finding) since the earliest days of radio.

Shields Up!

We need to make an important parting comment on the small loop antenna, before we move on. The role of the *shield* in receiving antennas has received a lot of misinformation over the years. Many high quality DF loops use electrostatic shielding, but for a very different reason than often cited. The loop (or loops, in the case of a multiturn antenna) may be enclosed in a copper or aluminum tube, with an air gap, generally in the center. Without the air gap, the shielding of the antenna would be complete, and wouldn't be able to pick up anything. The gapped shield reduces electrostatic coupling to the antenna, but it emphatically does *not* reduce noise pickup, contrary to some misguided press. What the shield *does* do is reduce or eliminate distortion of the antenna pattern, allowing it to perform its direction-finding tasks in the presence of metal reflectors, such as on a maritime vessel. Electrostatic shielding can also reduce the antenna effect making it less responsive to signals arriving from high angles. (The antenna effect is when a supposedly balanced loop actually acts as a very short whip. Since this "parasitic whip" is fairly non-directional, it will respond to signals arriving from very high angles, spoiling the normally predictable pattern of the loop.) The most effective DFing comes from signals arriving directly from the horizon.

Now, it is true that with a well-preserved pattern, you can achieve a greater null on some unwanted noise source, so in that regard, the shielding can indirectly reduce noise, but not in the manner often "advertised."

The small, tuned, shielded loop can be employed effectively as either a passive or an active antenna. However, the active version can result in even better balance than the classic passive device. We will describe a precision active DF loop in a subsequent chapter.

Chapter 9

Achieving the Perfect Null

As a general rule, achieving deep nulls in transmitting antenna patterns is not a high priority; in fact, deep nulls are often avoided intentionally. Having an unpredicted or unwanted "dead spot" in your transmitted wave can be highly undesirable, for what should be obvious reasons. However, there are two major reasons to strive for very deep or nearly infinite nulls in a receiving antenna. These are radio direction finding and interference reduction. Let us deal with the former first.

In Chapter 8, we dealt briefly with the *small loop antenna* as it is applied to radio direction finding. This is an important application of the small loop antenna, but more conventional antennas are indispensable in RDF work as well.

Nearly all high-performance direction-finding antennas rely on *nulling methods*. The reason for this is simple: it's a lot easier to achieve a very narrow and deep null in an antenna pattern than it is to create a very narrow beam. Narrow beams require lots of directivity and gain, and this translates into very large antennas. With careful design, it is possible to achieve a nearly perfect null on a receiving antenna array of nearly any size. In fact, at least in theory, it is always possible to design an antenna with an infinitely deep null in some direction. On the reciprocal side, it is impossible to build an antenna with infinite gain…even in theory.

Running Interference

One of the truisms of antenna theory is that *all* directional antennas rely on *wave interference*. Wave interference can be constructive or destructive. In the case of a *gain* antenna, constructive interference is the most useful phenomenon. In the case of a *nulling* antenna, destructive interference or *wave cancellation* is the important ingredient.

Let's examine perhaps the simplest nulling antenna, two dipoles out of phase. The two dipoles can be arranged in either a collinear **Figure 9.1** or broadside array **Figure 9.2**. The outputs of these two antennas are *added* out of phase. If the amplitudes of the two signals are identical, but

Figure 9.1 — Radiation pattern of two dipoles arranged in a collinear array in free space.

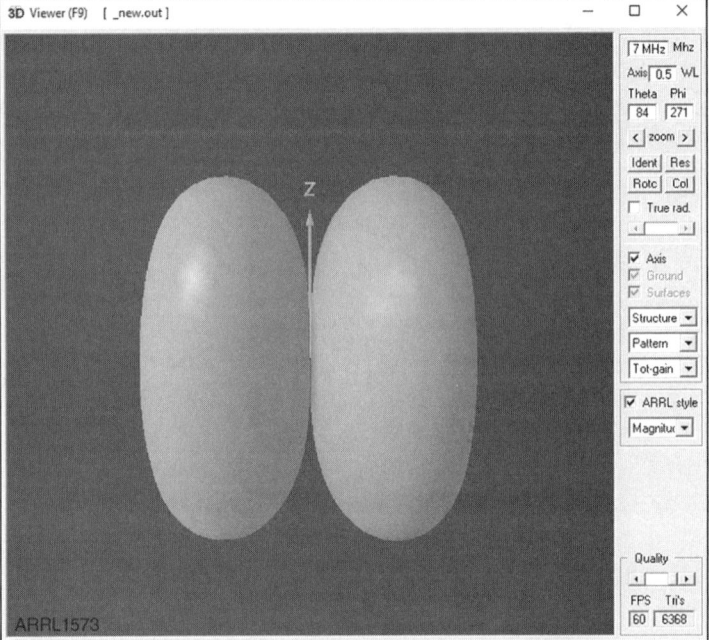

Figure 9.2 — Radiation pattern of two dipoles arranged in a broadside array in free space.

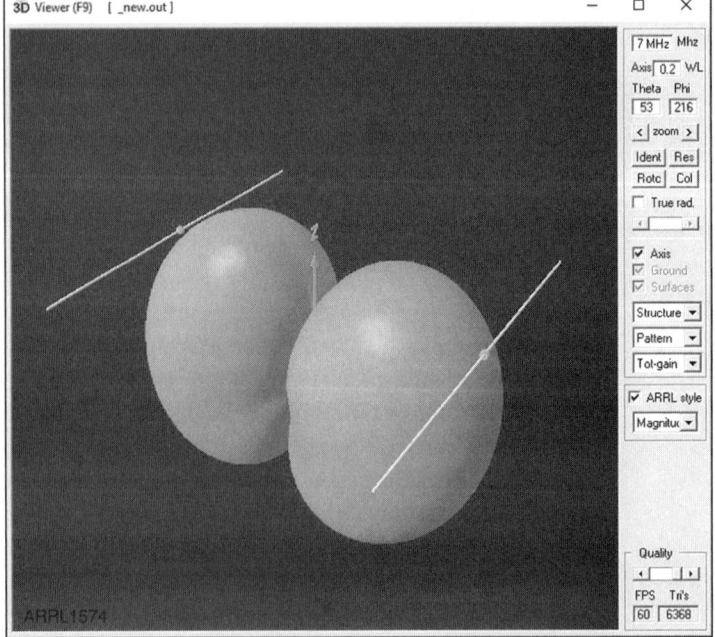

180° out of phase, the voltage at the summing point will be zero. This can only occur, naturally, when a wavefront intersects the dipoles at precisely the same time, which can only occur when the two dipoles are exactly equidistant from the signal source. The amplitudes will also, presumably, be identical, as any *practical* sized antenna will be very small relative to the total distance traveled by a wave in question. Any signal source lying on a line bisecting the segment between the two dipoles will be perfectly nulled out, regardless of the distance. This nulling function will work regardless of the spacing between the dipoles.

However, this antenna is also ambiguous, giving us two possible null directions, one being precisely 180° off from the true direction. Obviously only one of these can be right. This ambiguity exists whether the antennas are in a collinear arrangement or a broadside arrangement. What's a ham to do? If you're working with NVIS signals, this isn't too much of a problem, since you probably aren't going to be receiving many signals from directly underneath. The effects of real ground on Figure 9.1 and Figure 9.2 are shown in **Figure 9.3** and **Figure 9.4** respectively.

Figure 9.3 — The effect of real ground on the collinear dipole array of Figure 9.1.

Figure 9.4 — The effect of real ground on the broadside dipole array of Figure 9.2.

Making Sense of Antenna Sense

Fortunately, it is fairly simple to resolve the nulling ambiguity of a simple antenna array such as a pair of dipoles. The "right" and the "wrong" direction are a full 180° apart, which means you can eliminate the wrong one with an only moderately directional antenna. The most obvious solution to this dilemma is to have a second *unidirectional* antenna, to get you looking in the right general direction. Then you can use your extra-sharp nulling antenna for a precise heading reading.

In practice, it might not be convenient to lug around two antennas for simple direction finding. The preferred solution, in most cases, is to *temporarily* modify the pattern of a direction finding antenna to a unidirectional mode. A *sense antenna* is an antenna or antenna element which is combined with a nulling antenna to achieve a somewhat unidirectional pattern. Again, this does not need to be a very precise addition; you simply need enough directivity to eliminate all the possible *wrong* directions.

A sense antenna can be applied to any type of DF antenna, such as our anti-phased dipoles, or a more compact loop antenna. In general, a sense antenna pushes the overall antenna pattern into somewhat of a *cardioid* pattern (**Figure 9.5**).

Cardio Exercise

The major lobe of the cardioid pattern, achieved by means of a sense antenna, will be in the "correct" general direction of the transmitter in question. That is, if you design your sense antenna correctly! It is easily possible to tune a sense antenna so the *null* is toward the transmitter, too! How do you avoid falling into this rabbit hole? The most reliable method is to test your cardioid pattern on a *known* transmitter first. The second best option is to carefully adjust your component values based on a sense antenna design that someone else has already figured out. The ancient sense antenna designs in the *ARRL Antenna Book* are, fortunately, reliable….assuming you design them for the frequencies specified.

Now, the obvious question is, if you can get a really nice null with a cardioid pattern, why don't you just use *that* null for direction finding? The simple answer is, you *can*. The less simple answer is that it's a lot

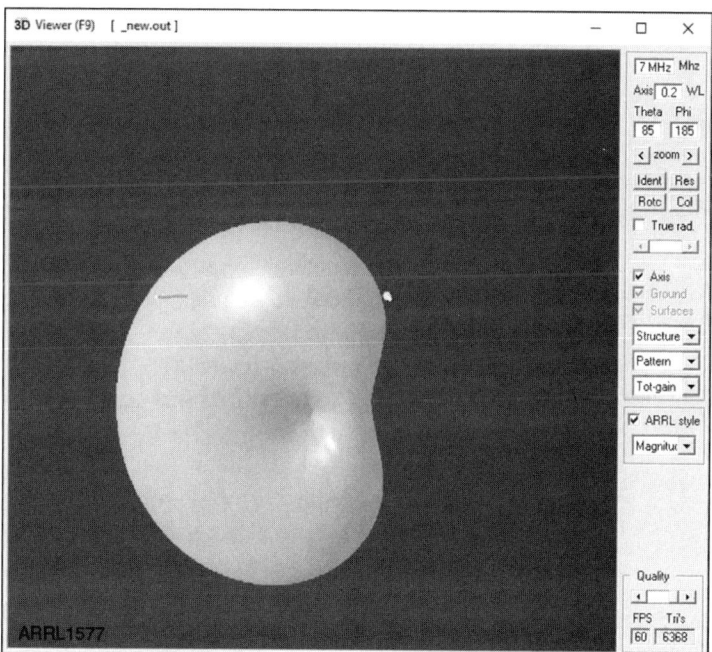

Figure 9.5 — Radiation pattern of a bidirectional RDF antenna with a sense antenna added.

more difficult to achieve a deep null cardioid pattern than it is to achieve a near perfect null with a symmetrical loop or dipole array.

The most direct way of achieving an ideal cardioid pattern with a pair of dipoles is with ¼-wave spacing and 90° *current* phase shift between them. The physical spacing is a trivial matter; while achieving the necessary 90° phase shift is extremely tricky, especially since there is a lot of mutual coupling between ¼ wave spaced elements. You cannot rely on 90° difference phasing lines, which will give you the correct *voltage* phase relationships, but not the correct *current* ones. The nice thing about the symmetrical dipoles with *180°* phasing is that any mutual coupling perfectly cancels out. You will *always* have a pair of perfect nulls. You just have to resolve the ambiguity. And that can be done sensibly with a sense antenna!

It should be mentioned briefly here that certain navigational systems *do* rely on well-designed cardioid patterns, such as VHF TACAN (tactical air navigation system). However, these are precisely engineered instruments, operating under tightly controlled environments. In most cases, you will find that the most effective amateur direction finding systems use some form of symmetrical array or loop.

Going Loopy

While the anti-phased dipole pair is the most direct and reliable means of achieving a perfect null, the next most practical solution is the *small loop* antenna. How do we define a *small* loop…other than that it's not a *big* loop? For our purposes, a small loop is one in which the current distribution is essentially equal around the entire circumference of the loop. This generally requires that the circumference of the loop is less than about ¹⁄₁₀ wavelength. There is no hard and fast line of demarcation between a small loop and a large loop; however the priorities and the properties of each are fairly distinct. A small loop achieves its end goals by very different means than, say, a full-wave resonant loop.

A full wave loop, either by itself, or as an element of a cubical quad antenna, for example, has maximum radiation *perpendicular* to the plane of the loop. And, as suggested, it has a well-defined resonant frequency.

On the other hand, the small loop antenna has maximum radiation *in* the plane of the loop, with very sharp, or even perfect nulls perpendicular to the plane. Although a *multiturn* small loop can operate as a resonant antenna (usually by means of an auxiliary capacitor), resonance is secondary to its operation, and has essentially zero effect on its directional properties.

Furthermore, a small loop, operated in its "normal" mode is very strongly vertically polarized. In most cases this strong polarization is an asset in its direction-finding abilities…but not always, as we shall discover.

Just Passing Through

The supreme simplicity of the small loop antenna belies the elegant physics of the device, to the extent that it's well worth the time to dive into this in some detail and ponder the life of an electromagnetic wave.

Let us consider a vertically polarized radio wave arriving from the bottom (**Figure 9.6**). The electric field intensity is represented by the amplitude of the arriving wavefront. Although this diagram *appears* to be showing a horizontal electric field, it is actually vertically polarized. We will ignore the magnetic field for now, which we have long since established as being perpendicular and in phase with the electric field. No need to clutter up the scenery with a bunch of redundant information.

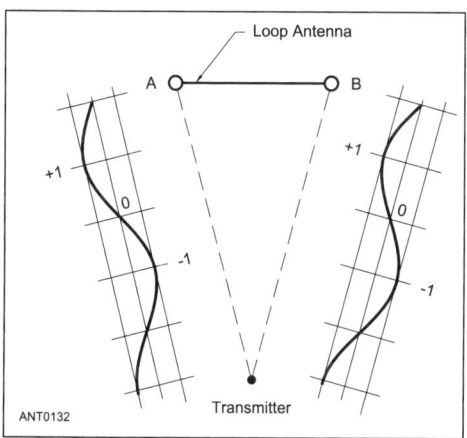

Figure 9.6 — A small loop shows a deep null for signals arriving perpendicularly to the plane of the loop. As would be expected, the loop shows a null for signals coming from the opposite direction, as well.

As we can see, the *instantaneous* amplitude of the wave, which represents instantaneous field voltage at any point, is dependent on its distance from the source. Of course, we also know that the free space attenuation is also dependent on distance, but we can ignore any differences in average field strength across any "normal" sized antenna. The voltage at any given point in space will fluctuate in length in a sinusoidal fashion as a function of time. But at any given *time*, the amplitude of the voltage is dependent on its *location*. This means that, at any particular instant, there will be a difference of potential between any two parts of the antenna that are at a different *distance* from the transmitter. In the first case, with the loop oriented broadside to the incoming wave, the voltage potential will be identical on the two vertical members of the loop. (It should also be fairly clear that the larger the diameter of the loop, the greater will be the difference of potential for any given frequency). This gives us a perfect null, just as in the case of two anti-phased dipoles…but for a very different reason. In the latter case, we have a null because of phase cancellation. In the former case, no voltage is induced in the first place!

Now, let's turn the antenna 90° (**Figure 9.7**) so that the incoming wave intersects both vertical sections of the loop at different times. (Again, while the amplitude wave is shown lying horizontally, this actually is vertically polarized.) In this case there will be a small but finite difference of potential between any two vertical sections, and thus some difference of potential at the antenna terminals. The small loop has a prominent role in lower regions

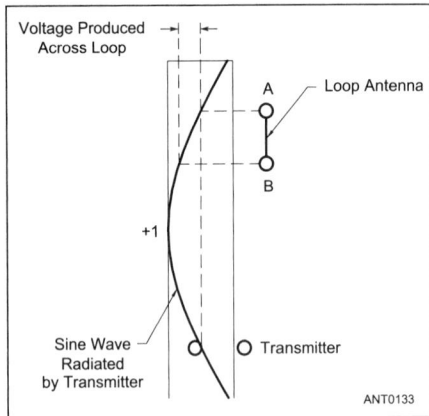

Figure 9.7 — Signals arriving edgewise to the small loop induce different voltages in the near and far vertical members. For angles other than 90 or 0 degrees, the output voltage of the antennas varies in a sinusoidal manner relative to the angle.

of HF clear down to ultra-low frequencies….down in the millihertz range! For Amateur Radio purposes, it is extremely useful as both a low-noise, and a direction finding antenna for the low bands. Again, this is a "low gain" antenna, where the priorities of low noise and directionality are far more important than sensitivity.

Non-Standard

The small loop antenna, as we mentioned above, is strongly vertically polarized. What happens if we attempt to receive a horizontally polarized signal with a small loop? If a horizontal wave intercepts the loop edgewise, there will clearly be no induced voltage, because the wires are at right angles to the electric field. So we definitely won't receive any signals, since simple cross-polarization is in effect.

But what about the case of a horizontal wave

Low Band Direction Finding for Fun and Profit

Direction finding is an important application of receiving antennas especially in the lower frequency amateur bands, but in recent years, low-band DF work has become somewhat of a lost art. "Foxhunts" on 80 meters were popular long before the activity became almost synonymous with 2 meters. The small (and very small) loop antenna reigns supreme in low-band DFing, primarily because of the very deep, well-defined nulls that such a loop provides. A sharp null is always easier to achieve than a very narrow lobe, regardless of the frequency. All commercial and military radio (and RADAR!) applications rely on nulling methods for precisely tracking radio sources, far and near.

There is no substitute for practice in the field for becoming a good low-band DFer, but the rewards (and fun) can be great, whether tracking down rogue radio interference, or simply winning your local foxhunt.

Keep in mind that, unlike the more familiar 2 meter foxhunt, low-band signals can arrive from several paths: direct, ground-wave, and skywave. To effectively DF an 80 meter signal, you have to know which of these paths the signal is taking. In addition, an HF foxhunt tends to cover a *lot* more real estate than your typical 2 meter hunt. Great patience, some understanding of propagation, and wide-area coordination are key ingredients.

Our two new low-frequency bands, 630 and 2200 meters, will present even more interesting challenges for direction finding activities. But undoubtedly, a small but dedicated community of hams will be up to the challenge.

coming in broadside to the loop? In this case, the "wave antenna" action would normally cause a null, because both *horizontal segments* of the loop are the same distance from the wave source, and would both have the same potential. However, since the antenna is still *fed* at the bottom, we now have a somewhat wimpy "normal mode" happening. In this case the loop looks somewhat like a folded dipole…though a very small one. The bottom line is that the small loop antenna is a very poor direction finding antenna if the polarization is not correct.

Another thing that can spoil the performance of the small loop is signals arriving from high angles. While the direction finding properties of the small loop are excellent for vertically polarized ground waves, high angle sky waves are a different story…especially since you no longer know what their polarization is to start with. At low or zero vertical angles — true ground wave — radio signals *must* be vertically polarized. This is not true for signals coming straight down…or nearly so, such as with NVIS propagation. The takeaway here is that, the small loop antenna is an excellent direction finding antenna *under the right conditions.*

Just for Fun

If you have a portable transistor AM radio, you can demonstrate just how *good* a loop antenna is under the right conditions. The small loop antenna inside a typical transistor radio is actually a ferrite "loopstick" antenna, but it has the identical radiation pattern of a larger air-core small loop. A ferrite loopstick has a perfect null perpendicular to the plane of the windings…which happens to be *in line* with the ferrite rod.

I live about 2 miles from a 50,000 W AM radio station at 1170 kHz (with a half-wave tower). With my el cheapo transistor radio, I can still achieve an absolutely perfect null of that powerful station by carefully steering and tilting the thing. It's unbelievably sharp…but it's unbelievably deep, too! If I use my safecracker touch, I can make the radio station completely disappear.

If you have the opportunity, you can try this same experiment. Try it on a number of local stations, and then some distant ones late at night. You'll find you can usually get a perfect null on powerful ground wave stations, but nowhere near as good a null on distant skywave stations! This is because the distant stations will be arriving at high angles with non-ideal polarization for the loopstick.

A number of small tuned ferrite loop antennas are available for the radio amateur for both direction finding and general reception. The better ones have both a precise azimuth steering mechanism as well as a vertical angle adjust. Now, why would you need a vertical angle adjustment? The main reason for this is that a vertically polarized ground wave signal will *tilt* slightly in the direction of its travel. If you can tilt the angle of your ferrite rod to match this, you have a better chance of achieving the perfect null you so passionately desire. We'll talk about this wave tilt in more detail as we explore the Beverage antenna in a later chapter.

Reducing Antenna Effect

It might seem a bit odd to label something like "antenna effect" as a *defect,* especially in a book about antennas! There are actually three names (that I know of) that describe the same deficiency. Really old texts refer to the same thing as "night effect" and some semi-old texts refer to it as *skywave effect.* But they're all the same thing. A conventional direction-finding loop is only on its best behavior for signals arriving from very low vertical angles, ideally at *zero* degrees above the horizon. At other angles, the null of a small loop becomes less defined, or under extreme conditions can go away entirely.

A "normal" DF loop works best during broad daylight when there is no ionospheric path. In this case, the *ground wave* is the only wave one needs to contend with, and very predictable and deep nulls are available. This can be easily demonstrated with a cheap portable AM radio while listening to a local AM broadcast station. You can orient the radio to obtain a perfect null. I always demonstrate this to my astonished students by nulling a powerful 50,000 W AM station less than two miles away. With proper orientation of the receiver's loopstick, the station just *goes away* — entirely.

This is not the case at night, when ground wave signals may be accompanied by skywave signals, or in situations where there are *only* skywave signals. In some cases it *may* be able to perfectly null a high angle skywave signal by *tilting* the axis of a small loop to match. In fact, there are some commercially made loopstick antennas that have both azimuth and elevation adjustments for just this purpose.

Now, unless the loop is perfectly balanced, that is it acts *purely* as an inductively coupled loop, it can also act as a very short whip antenna. This "parasitic whip" of the unshielded loop can greatly modify the loop's directional pattern. (This is sometimes done *intentionally* by means of a *sense antenna*, as described earlier. But under actual operation the sense antenna is switched out. You do not want an *unintentional* sense antenna to exist). A well designed, electrostatically shielded loop can greatly reduce this antenna effect, eliminating the "parasitic whip" characteristics of the antenna.

However, even a perfectly balanced, shielded loop can suffer the effects of high-angle signals. In fact, most commercial literature recommends *not* using the small loop antenna when sky waves are likely to exist. Instead, sky wave suppressing antennas, such as the Adcock antenna, are recommended for direction finding work with non-ground wave signals. An upgraded Adcock antenna is described in Chapter 22.

Hybrid "loop-like" Adcock variations are also very effective, allowing lower frequency operation than would be practical with the classic Adcock antenna.

You *Can* Make Accurate Field Strength Measurements

This is a good time and place to review the principles of the Scientific Method. A great deal of the hogwash, balderdash, and flooby-dust surrounding antenna performance would go away if we would occasionally touch base with good scientific and engineering practice.

There is little point in developing and experimenting with antennas if we have no reliable means of measuring the results of our efforts. Taking a brief detour into the realm of *transmitting* antennas, isn't it safe to say that we should evaluate the performance of said antennas by actually measuring field strength at a distant location? Is not the end goal of all our station improvements to *optimize* the signal strength at the target location?

When it comes to antennas, most hams are inexplicably obsessive about measuring and tweaking *intermediate processes* (such as SWR), while totally disregarding the end product.

In many cases it is not necessary to make precise absolute measurements, but it is *always* beneficial to be able to make meaningful *relative* measurements. In other words, does twiddling element A *improve* or *deteriorate* signal strength at location B?

As one simple glaring example, for decades and decades hams have debated whether a cubical quad antenna has more or less gain than a Yagi antenna (presumably under the same conditions). Such pointless arguments could be settled once and for all with a simple *relative* field strength reading at a distant location. The lack of actual measurement can go a long way toward propagating endless discussion, but it does little toward technological improvement.

Many hams will argue that *there are too many variables* to make field strength readings of HF antennas. This assertion is just lame science. Whether you're dealing with chemistry or physics, or psychology, part of good laboratory procedure is *establishing* the test conditions, placing all the test subjects on a level playing field. Sure there are a lot of variables when it comes to HF propagation…but if you specify the same time and location for each specimen, all those variables cancel out. We do this sort of benchmarking in every other science. What makes antenna measurement so "special" that we feel we can invalidate any meaningful measurements?

Now, let's redirect our attention back to receiving antennas. When you *know* how your receiving antenna is performing, it not only helps *you* but it helps every other ham who happens to contact you. Of course, one of the most important things you can do to this end is to give accurate signal strength reports. But now, all the objections start flying. "S-meter readings are meaningless anyway, so we just give everyone an S-9…at least in a contest." While we have already established that the S-meter reading is *not* the ultimate goal for receiving purposes, but rather the signal-to-noise ratio, there is still a great justification for giving reliable S-meter readings. Usually this is for the *other* guy's benefit. It's also very useful for performing a lot of radio science measurements.

But, in addition, you need to ask yourself; *why* is my S-meter meaningless…other than the fact that everyone else says it is? Here are three lame excuses to disregard S-meter readings, and the correct associated responses.

1) I have no idea what my receiving antenna gain is.

Aren't you at least a little curious? Is your antenna better or worse than a dipole? If it is a dipole, wouldn't you at least like to know if it's performing about as well as a dipole should? Do you have any idea what your antenna pattern is? Do you even know if your major lobe (or lobes) are in the right general direction for the location of interest?

2) The S-meter on my receiver or transceiver isn't calibrated to anything.

This was a good excuse in 1960. Many modern receivers have good S-meter calibration across the HF range, and if they aren't already calibrated, it's a simple process to do so. A great deal of ingenuity went into the design of the original S-meter. It is a shame not to use it to its full advantage…or come up with something even better.

3) I have no idea what the HF path loss is. Isn't the dynamic range of the ionosphere something like 200 dB or more? How can you make any meaningful signal strength reports under such variable conditions?

Isn't one of the reasons we do ham radio is to *figure out* things like propagation path loss…and the contribution to radio science it facilitates (and who knows what else)? If you know the station performance at both ends of a radio path, you can learn an awful lot about the intervening path and the associated physics.

Again, while achieving maximum SNR at the receiver is the primary goal for practical operating, there are some good reasons for knowing the signal strength as well. Knowing the performance of a receiver/antenna combination is important for measuring transmitting performance, if only indirectly.

Additionally, in the commercial radio field, it's often necessary to make accurate field strength measurements. I spent nearly a quarter of a century taking field strength measurements for an AM broadcast station in subzero temperatures during half the year, and slapping swarms of mosquitoes during the other half. A typical calibrated and very accurate AM broadcast field strength meter is shown in **Figure 10.1**. While such an instrument is seldom necessary for "normal" Amateur Radio applications, knowing how to use such an instrument is extremely useful. Without

Figure 10.1 — A typical calibrated and very accurate AM broadcast field strength meter. The hinged case cover (above the meter front panel in this photo) includes a shielded loop antenna.

proper measurement methods, it's impossible for us to "advance the state of the radio art," which should be every radio amateur's goal…or at least one of them.

A Few Simple Tricks…and a Project

The bottom line in this is that with a little determination, you *can* make meaningful and accurate field strength readings, across the entire Amateur Radio spectrum. However, for our purposes, we will concentrate mainly on the HF and MF regions, where you are most likely to be using the types of antennas described in this book.

As we discussed earlier, *relative* or *comparative* measurements can be extremely useful…and in fact, for most day to day operation, relative measurements are the most important. If you're trying to steer your antenna for best reception of a distant station, for instance, a relative signal strength reading is more than adequate — you simply steer for the greatest intelligibility. This may or may not coincide with the highest S-meter reading, but it's usually a fairly practical starting point.

Let's look at something a bit less obvious, now. Let's say you have an HF transceiver with a built-in SWR meter, an external manual antenna tuner with its own SWR meter, and some kind of transmitting antenna out in the back forty. Now, as you tune everything up on your favorite frequency, you notice that there is a bit of a discrepancy between the SWR meter in your transceiver and the SWR meter in your antenna tuner. The former says 2:1 SWR and the latter says 1:1. (The reasons for this can be numerous, but we won't discuss those here.) The important question is: which one do you trust? Or do you trust neither?

To answer this question, you need to ask yourself what you're trying to accomplish in the first place, with all your tuning and twiddling. (A surprising number of hams fuss and fume and sputter and sweat when asked this simple question.) *Presumably*, your goal will be to achieve the most effective communication with the station at the far end. Yes, achieving a low SWR is probably nice at some point in the process, but it's not the end goal. Your goal is to make radio communications work.

Since you now have two questionable SWR meters, how should you adjust things to achieve your desired end goal — maximum signal strength? Would it be too much of a stretch of logic to consider measuring the actual signal strength?

Now, ideally, you'd want to measure the signal strength at the distant station. However, if you're still tuning things up, there most likely won't *be* an operator at a distant station to give you any feedback. (There are, however, some automated tools available — see the sidebar, "Digital

Game Changers.") The next best thing is to measure the signal strength near your transmitting antenna…an extremely simple thing to do in most cases. Since we're only shooting for maximum signal strength — and not attempting to come up with any actual numbers — the correct solution would be to adjust your antenna tuner for the maximum signal strength. Now, if this condition happens to coincide with a 1:1 SWR reading on one of your SWR meters, that's even better.

The sad fact of life is that very few hams implement any means of measuring radiated field strength, whatsoever…even if they have thousands of dollars invested in other instrumentation. There's very little excuse for this.

The bottom line is that accurate field strengths measurements can be made with readily available amateur apparatus. All that is missing is a determination to do so.

Digital Game Changers

Some recent developments in HF radio have made evaluation of Amateur Radio signals much more believable. The Reverse Beacon Network (RBN) is one such tool. With the advent of the "Internet of Things," it is now a simple matter to have a radio receiver controlled from across the room, or across the nation. The Reverse Beacon Network is a network of internet accessed radio receivers that allow hams to monitor their own signal levels from afar. Whereas, in the past, continual monitoring of radio propagation was pretty much relegated to WWV/WWVH and shortwave broadcast stations, the RBN can supply essentially unlimited monitoring capabilities.

In addition to the RBN, several recent digital HF modes have built-in signal strength and signal-to-noise ratio reporting. WSPR, the Weak Signal Propagation Reporter, is one of the more elegant of these protocols. WSPR beacons can be built very inexpensively, and are a great project to get your feet wet in homebrewing. Other weak signal modes such as FT8, JT65, and JT9 (part of the WSJT-X package) require no hardware that you probably don't already have, namely a computer with a sound card, and an HF radio. The WSJT modes are specifically designed to allow not just weak signal communications, but meaningful measurements of signal-to-noise ratio, allowing hams to do real radio science on a shoestring budget.

As of this writing, we are in a period of deteriorating radio propagation in the higher HF bands, and such "newfangled" technologies will allow us to eke out the last remaining vestiges of HF propagation, at least for the remainder of this sunspot cycle. In addition, these methods should prove very useful for our two newest bands on 630 and 2200 meters.

It should be noted that *no* digital technology is a substitute for good RF engineering and design. Low-noise receiving antenna methods *and* the latest digital methodology will allow hams to continue operating under conditions once deemed impossible.

Just a Sampling

If you desire to measure the voltage across some component in an electrical circuit, you would typically touch two probes of a voltmeter across the component in question. The voltage you'd read is the *difference of potential* between the two points. An electromagnetic wave traveling through space doesn't really have any "terminals," so a different way of viewing the problem comes into play.

In order to take a sampling of field strength, we need to "immerse" our test equipment into the field somehow. The standard "immersion" tool is a length of wire, typically 1 meter long. When such a wire is inserted into the electromagnetic wave, a potential is induced along the wire, proportional to the electric field of the electromagnetic wave. There will be a potential between the two ends of the wire, which *can* now be measured with some kind of voltmeter.

Microvolts per meter (μV/m) is a convenient standard for measuring field strength. A field strength that induces 1 μV of potential along 1 meter wire (of the correct polarization, of course) is, naturally, a field strength of 1 μV/m. For practical reasons, it's more convenient to cut the reference antenna in the middle and measure the voltage across the gap than to measure the voltage potential between the far ends. It should be obvious that to attempt to do the latter, you would create something like a folded dipole, composed of your sample antenna and your test leads.

There's nothing particularly sacred about a center-fed dipole, either for transmitting or receiving, but it's a convenience when performing measurements. A standard "reference gain" antenna, therefore, may or may not be actually 1 meter long, but is designed so that a standard field (of say 1 μV/m) results in 1 μV appearing across the terminals, assuming an open circuit at said terminals.

A true field strength meter would absorb no actual power from the "ether." As is the case with any measuring instrument, we don't want the instrument to actually *affect* what we're measuring. This is why we want voltmeters and oscilloscopes to have extremely high input impedance. In reality, however, a field strength meter is a receiver, and like any receiver, presents a finite load impedance to the antenna fixed across its input terminals. For precision, it's easier to work with a known, fixed load impedance than to attempt to create a true infinite impedance load for your sample antenna. For the purposes of measurement, we aren't trying to optimize anything; we're just trying to report on the facts as they exist!

Attenuators

One of the common arguments about using a receiver's S-meter as an actual instrument is that the linearity, accuracy, and general reliability of an S-meter is always questionable. While we would tend to disagree with such a blanket condemnation of the S-meter, we can offer a method that will all but eliminate the vagaries of the typical S-meter. This method is to use a precision attenuator (which is perfectly linear by its nature) ahead of the receiver, while using a fixed S-meter reading as a reference. This is especially meaningful and accurate when comparing different antennas on the same frequency, or determining the radiation pattern of a given antenna on a given frequency. Several useful step attenuator projects have been described in the *ARRL Antenna Book*, and past *QST* articles.

Most of these attenuators use mechanical switches for inserting the desired amount of resistance (usually in the form of H or T pads), but more advanced ones can use PIN diodes or other solid-state switches. This latter enhancement is useful when applied as an automatic gain control on a receiver that has no "inherent" AGC, such as a typical direct conversion receiver. The real advantage of using an attenuator instead of a gain-controlled transistor or IC stage is that the dynamic range is not compromised.

We will conclude this chapter with a simple but versatile relative field strength meter, using the AD8067 op-amp, featured several other places in this book, and the extremely useful AD8307 log amp. See the sidebar "Logarithmic Amplifiers — Smooth Operators in the Digital Age" at the end of this chapter.

A Most Useful Shack Accessory

There was a time when just about every ham had some sort of field strength meter in the shack. For a number of reasons, these simple accessories have fallen out of favor. While, as we've mentioned earlier, accurate *absolute* field strength measurement can be difficult and expensive, it is a much simpler matter to measure *relative* field strength. Being able to measure *changes* in radiated field, either close in or distant, is what you need when performing antenna adjustments…or even transmitter tuning. There's really no excuse for a ham not to have some means of measuring radiated field strength, if only to confirm that his or her transmitter is actually functioning!

Over the years a number of inexpensive field strength meters have been available, often incorporated with an SWR meter. These are generally *passive* devices, using a simple diode detector and a microammeter.

Figure 10.2 — A homebrew field strength meter using a steel shish-kebob skewer for the whip antenna.

The sensitivity of these devices leaves something to be desired.

Here we present a slightly more useful instrument, consisting of a sensitive active antenna, using the AD8067 in a non-inverting configuration and an AD8307 logarithmic amplifier (log amp). The unobtrusive little box is shown in **Figure 10.2**. (Yes, that's a steel shish-kebob skewer used for the whip antenna).

The sensitivity of this meter is frequency dependent, as is the case with all such simple instruments. While, with a great deal of effort, it *is* possible to build a compensation network to make the sensitivity uniform across the entire HF spectrum, there really isn't a great need for this kind of complexity for a *relative* field strength meter. The AD8307 logarithmic amplifier serves as the actual detector, as well as the "linearizer." **Figure 10.3** and **Figure 10.4** show the complete schematic diagram. With a uniform scale meter, such as a typical milliammeter, each division will represent a fixed number of *decibels*. The overall circuit has a tremendous amount of dynamic range. While you probably won't be able to measure the field strength of that rare DX station, you *will* be able to use it to measure relative local field strength of any imaginable QRP or QRO station.

The layout is not critical, though you should use good RF design techniques. As described earlier, the AD8067 is only available in surface

Figure 10.3 — Schematic diagram of the active antenna section of the field strength meter using the AD8067 op-amp in a non-inverting configuration.

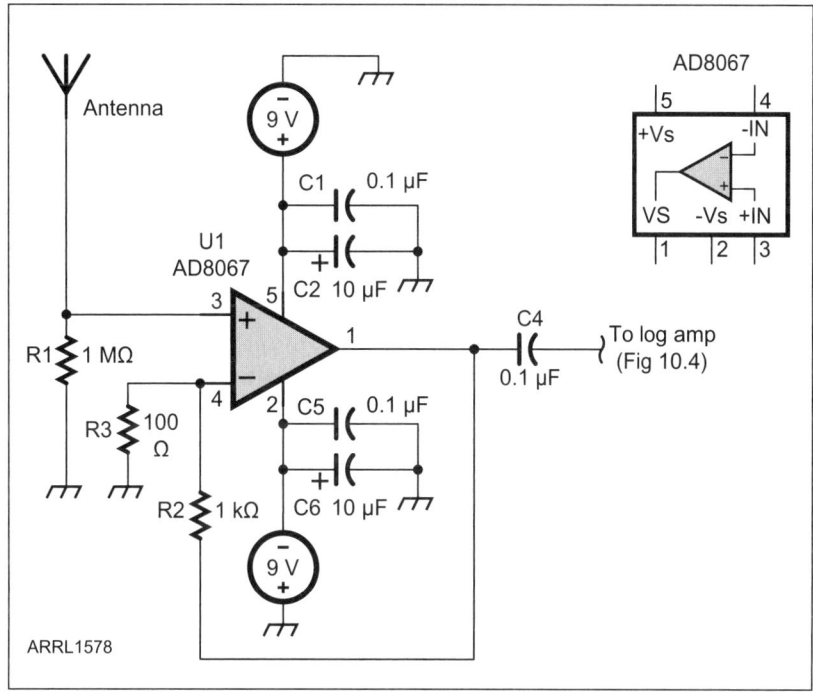

Figure 10.4 — Schematic diagram of the logarithmic amplifier section of the field strength meter using AD8307 log amp.

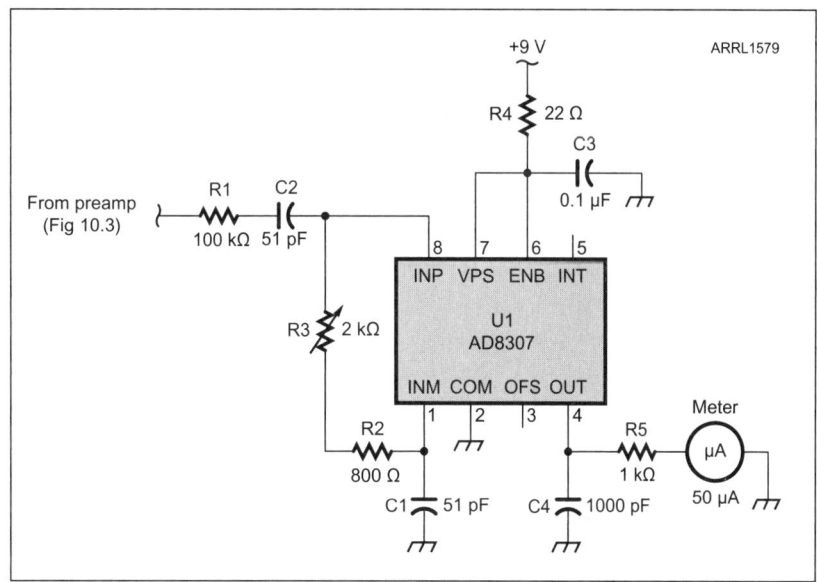

mount form, so you will either have to use an adapter "daughter board" to allow through-hole assembly (as I have done in this specimen), or etch a complete surface mount board suitable for the chip. As far as the AD8307 log amp is concerned, while the datasheet recommends entirely shielding the device, I haven't seen any problems at HF using "normal" construction. What I like about this chip is that there are only six pin connections when used in this application. The log amp *is* such a versatile device, with countless applications in the ham shack, that you might want to explore it further by downloading its datasheet from the Analog Devices website (**www.analog.com**). The field strength meter in question is only the *simplest* application of this wonderful technology.

Since the field strength meter *is* an active antenna, the physical connection of the whip *must* be made to the preamplifier with a very low capacitance connection. This precludes any form of coaxial cable, or most other transmission lines. Use a short wire from the bottom of the whip to the non-inverting input of the AD8067. The length of the whip is non-critical, though, of course, the sensitivity will increase with added length.

Shameless Commercial Plug

Our little company, AlasKit, will in the not-too-distant future, be producing a clever variation of this field strength meter in a metronome shaped cabinet, with a vertical LED bar graph display. The tentative name for this is the "Fetronome" in deference to the FET op amp. (I wish I was clever enough to come up with that name, but one of my partners-in-crime suggested it. I think it might stick, however. Stay tuned.)

Logarithmic Amplifiers — Smooth Operators in the Digital Age

Nearly every day, something or someone reminds me that I'm getting older. I'm not complaining, mind you. If you wait around long enough, you can dredge up a lot of ancient, forgotten methods and hardware and introduce them as your own discoveries. This way you can come across as the repository of all human wisdom to the young whippersnappers crouching at your ankles. I do this frequently in the electronics classes I teach, eliciting spontaneous rounds of *ooohs* and *aaahs* from my rapt audience.*

Steam Punks and Slide Rules

Once in a very great while, something that comes from popular culture is actually beneficial to human intelligence. Because of "Warehouse 13" and the "Steampunk" subculture it helped generate, there has been a huge renewal of interest in older technology, including slide rules (and other mechanical calculators). These have become extremely in-demand, high dollar items, which could be a great financial boon to me personally — if I had any interest in selling them — which I don't. Maybe people are beginning to worry about how they'd ever make change at the local Stop 'n' Go in case someone ever "pulls the big plug." In any case, I'm finding more people interested in how a slide rule works than I did a decade ago.

This semester, I have one student who is even more grizzled than I am. In fact, he could easily teach my class. One day he brought in his collection of slide rules, and I brought in mine, and we spent a good half hour bewildering our unenlightened comrades as we compared our collections. I used this "teachable moment" to demonstrate how one could perform just about any calculation at will by the judicious use of logarithms, an application where the slide rule excels.

Hello Operator

Most hams have at least a passing familiarity with the operational amplifier, or op-amp. These are generally viewed as convenient black boxes for producing lots of amplification with just a few parts. However, most hams are fairly oblivious to the original intent of the op-amp, and that is for performing *mathematical operations.* It's why it's called an op-amp in the first place. They can perform such functions as multiplication, division, squaring, taking square roots, and even simulating (modeling) physical phenomena such as RC time constants.

Off the Straight and Narrow

Although op-amps are typically noted for their extremely linear amplification properties, this was *not* the original selling point. In fact it was their controlled *nonlinear* behavior which gave them their tremendous usefulness!

Let's take a look at a fairly simple, but not too simple, nonlinear mathematical operation: squaring a number. A simple diode will do this quite nicely — over a very limited range. There is a region in the forward conduction range of a diode called the "square law" region. In this region, somewhat less than a quarter of a volt, the *current* through a diode is equal to the *square* of the applied voltage. At voltages much above this, the diode becomes nearly linear; it acts like a simple resistor. This is called the "boring region." No, actually, it's called the linear region, but it is rather boring, nonetheless.

We should note here that a vacuum tube diode also has a non-linear region like this, but it follows the "3/2 power law" instead of the square law. Unlike the semiconductor diode, the 3/2 power law is not limited to a tiny fraction of the operation range of the vacuum tube diode, but is in operation at all times. The problem is, the 3/2 power law is not as useful a function as the square (or power of two) law. But it's not entirely useless either, as we shall discover later.

Because the square law region of a semiconductor diode *is* so limited, we say that the diode square law region has a small *dynamic range*. It can only perform its desired mathematical operation over a very small range of input numbers. (At least not without some fancy footwork ahead of time.) This fancy footwork, which we will discuss in depth shortly, is known as "prescaling." But first we need to fill in a few blanks.

The Universe is Analog

Contrary to what you may have been told, the universe doesn't come in discrete chunks (except at, perhaps, the most microscopic levels). Values such as distance, time, mass, temperature, and speed come in a *continuum* of values. Your car doesn't instantaneously jump from 10 miles per hour to 30 miles per hour. (Well, actually, I had a Pontiac with a bad transmission that would occasionally do this, but it's not the norm.) In the real universe, you pass through an infinite number of speeds in passing from 10 to 30. This is true of *any* physical value you may encounter.

One of the problems with mathematics is that we confuse numbers with the things the numbers represent. A number is not a physical value; it's simply a way of *expressing* a physical value in a convenient manner. The proliferation of digital electronics (and digital everything else) further masks the physical reality we live in. There's nothing inherently wrong with digital electronics or digital processing. In fact, I was one of the earliest users of software defined radios. The problem is that the typical digital designer is like the man with a hammer...everything he sees looks like a nail.

Now, just as we can use *numbers* to represent *physical values*, we can use *physical values* to represent *numbers*. Voltage is the most convenient method of representing numbers, and that is what we do in *analog computers,* of which the logarithmic amplifier (log amp) is just one sort.

One way we can represent numbers as voltages is with a simple linear *coding* scheme, each number from 1 to 10, for example, being represented by a voltage from 1 to 10 V. In fact, we can use a simple *summing* amplifier (just an op-amp with two input resistors) to add two voltages (representing their matching numeric values) and arrive with the mathematical sum at the output.

However, this scheme runs into problems if we try to use our square law diode. It only will square numbers between about 0.01 V and around 0.2 V. Obviously a 1 through 10 V input range is going to be pretty much outside the range of our "squaring" computer.

To keep our input numbers (voltages) within the range where our computer will compute, we need a device called a *prescaler*, which simply converts any input numbers we need to deal with into values inside the magical square law range. Simple prescalers can be made of resistive voltage dividers. Now, the only problem with prescaling our input numbers to make our diode computer happy is that the answer will also be scaled down. So we have to re-amplify the output voltage (answer) by the same amount we scaled it down. This process is called *normalizing.*

Prescaled and normalized squaring amplifiers (also known as exponential amplifiers) were very common before the age of the digital computer. Unfortunately, for general purpose computing this took a lot of high-precision components.

Rather than merely accommodating the limited dynamic range of the diode computer with prescaling and normalization, it's actually possible to expand the dynamic range of the diode square law region by incorporating that diode in the *negative feedback loop* of an op-amp. This actually does something else for us, too; it converts the *exponential amplifier* into a *logarithmic* amplifier, which now gives us the rudiments of an electronic slide rule. An op amp with a diode in the feedback loop gives us an output voltage that is the logarithm of the applied input voltage.

As Easy as Falling off a Log

A slide rule multiplies numbers by adding the logarithms of those numbers. If we take two logarithmic amplifiers and follow them with a summing amplifier, we now have an electronic multiplier. Now those of you with a few years under your belt might protest: "But just a simple diode is a multiplier, too! Why all this extra gobbledygook?"

Yes...a lonesome diode is indeed a multiplier. And a squarer. And a "cuber." And an adder. And a subtractor. Unfortunately, it does all of these (and more) all at once! As you might imagine, this may not be mathematically optimal. And, as in the previous discussion, it has limited dynamic range. On the other hand, a proper analog multiplier will perform *just* the operation you need — and with a very wide dynamic range.

We can also perform division by subtracting logarithms. This simply requires that we apply the outputs of our log amps into *differential* inputs of our summing amplifier. Pretty slick!

Order! Order!

We can cascade log amps or exponential amps to achieve higher order mathematical functions. For example, if we cascade two "squaring" amplifiers, we have a fourth power amplifier. Or, we can cascade two square root (log amps) to achieve a 1/4 power function. (By the way, this particular function is very useful for radar AGC systems, where the received signal generally decreases as the fourth power of the distance to the target. This is from the standard radar equation.) Radar people like log amps. As do astronomers. In fact, any scientific application where one is likely to encounter widely ranging signal levels of any sort is likely to be using a few log amps. Or a lot of them.

*Nearly spontaneous, that is. Actually, on the first day of class I coach my students to say Oooh and Aaah in unison whenever I demonstrate a phenomenon of profound interest. I have prepared cue cards for prompting this behavior. This is a great boost to my ego, and it keeps the aeronautical ground school class across the hall continually guessing what we're up to.

Chapter 11

The Aperiodic Loop

Lumpy and Constant

When speaking of antennas and antenna theory, it's common to differentiate between these things called "lumped constants" and "distributed" constants. In a "normal" electrical circuit, we consider electrical components, such as resistance, inductance, or capacitance, all being confined to a specific location, a point in space. We have a lump of resistance, a lump of inductance, or a lump of capacitance. We don't normally consider the time it takes for an electron to travel through a resistor; everything that happens at one end of the component happens simultaneously at the other end.

What makes an antenna different from a "normal" circuit, indeed the thing that allows it to function at all, is the fact that there is a spatial element involved. Things that happen at one end of an antenna don't happen at the same time as things that happen at the other end. Resistance, inductance, and capacitance are distributed along the entire length of the antenna. So, the question naturally arises, at what point does a circuit component change from a lumped constant to a distributed component?

As it turns out, every component is a little bit of both. No matter how small you might build an inductor, for example, there is always going to be *some* radiation happening — which is a distributed property — as well as pure self-inductance, which is very much a "lumpy" phenomenon.

Harmonic Antennas

There is a term used in older radio literature that you don't see much today. The term is "harmonic antenna." Probably the reason for this is that most antennas we are familiar with *are* harmonic antennas. They resonate at a specific frequency (and multiples thereof). Perhaps this is the *most* telling way we can differentiate between a lumped constant circuit and a distributed one. A lumped constant circuit consisting of a resistor, an inductor, and a capacitor has one and only one resonant frequency. (This assumes, of course, that each component is a pure component, not containing any "parasitic" elements. All inductors have some resistance, all resistors have some inductance, and so forth. But for most practical purposes, we can ignore those parasitics.) On the other hand, a distributed circuit such as antenna has multiple resonant frequencies — the fundamental, plus multiples or harmonics.

It is the capability of suspending harmonics that gives the "normal" antenna its *periodic* nature. It has an impedance that periodically repeats...at least to some degree. Both loops and dipoles exhibit periodic behavior.

A Definition or Two

In contrast with the *harmonic antenna*, the *aperiodic* antenna can be defined as one operated well below the first harmonic (fundamental frequency). The simplest example is the very short whip, which dominates the first part of this book. The second most common example is the very small loop. However, in modern nomenclature, the *aperiodic loop* is generally considered to be a small loop with resistive loading, which further reduces its periodicity and provides other useful benefits.

A number of interesting and useful receiving antennas have shown up in recent years. A happy coincidence with this development is the wide acceptance of antenna modeling methods by a large number of hams. Many of these recent receiving antennas operate in a somewhat unintuitive manner, and would be nearly impossible, or at least extremely time-consuming, to design using old-school cut-and-try methods.

While we need to reiterate our caution about using antenna modeling as a substitution for thought, we need to acknowledge that modern antenna modeling programs are excellent when used within their limitations. With a few exceptions, most properly modeled antennas actually do perform as advertised when actually constructed! At least at HF frequencies, large departures between modeled antenna performance and actual antenna performance are usually a result of inadequately modeled *ground* conditions. Actual ground conditions can vary immensely over small geographical distances, and so the "silver bullet" ground model does not exist. Modeling an antenna in free space is entirely independent from measuring the effects of ground (as long as there is no mutual coupling to ground), so free space performance is usually most meaningful. With that one caveat in mind, we can generally, with great confidence, use antenna modeling as an integral part of Amateur Radio antenna design and deployment.

One example of where antenna modeling is almost, if not an absolute necessity, is in the design of the aperiodic loop. In this writer's opinion, the aperiodic loop is one of the more intriguing antennas in existence. While this term actually refers to an entire *class* of receiving antennas, the *traditional* aperiodic loop is fairly standard. For many years this antenna has been used in scientific and military applications, but is only recently being discovered in Amateur Radio circles. It has been given a passing mention in the *ARRL Antenna Book* for a number of years, but not widely adopted by hams because of a lack of meaningful construction details available. We will describe a variation of the classic aperiodic loop designed by Michael Trimpi, emeritus, of Dartmouth College. This variation, which we will call the "Trimpi Loop" is widely used in ionospheric research.

Aperiodic loops are, almost without exception, deployed as *active* antennas. However, unlike the standard active whip antenna (also an aperiodic antenna), the aperiodic loop, lacking any resistive loading, has a *very low* source impedance. The requisite associated preamplifier must therefore accommodate this different impedance. There are two general ways of approaching this problem. One is to use an impedance matching transformer, and the other is to use an amplifier with an inherently low value of input impedance. The common-base amplifier is one common configuration of the latter. It has many similarities with the grounded grid vacuum tube amplifier, such as wide bandwidth, low input impedance and inherent stability. Both balanced and unbalanced common base amplifiers have been used for many years in conjunction with small loops especially on the low and very low frequencies.

I have included Michael Trimpi's description of his loop in its original form in the Appendix at the end of this book. As with most meaningful innovations in radio, the "Trimpi Loop" is deceptively simple. This configuration uses a step-up transformer to an op-amp with a typical high impedance input. A number of useful and interesting equations relative to the very small loop are shown in Trimpi's paper. Of particular interest is that the induced voltage is proportional to the enclosed *area* of the loop (that is also true for the resonated small loop).

Interestingly enough, the *gain* of a full size resonant loop (such as used in a cubical quad) is also proportional to the enclosed area of the loop. The salient point here is that there is always a price to be paid in making *any* kind of antenna smaller, but the tradeoffs are usually worth it in terms of signal-to-noise ratio, or in bandwidth (as in this particular case), which are higher priorities than straight gain, when it comes to receiving antennas.

The especially clever design of this antenna is the way the residual inductive reactance of the antenna is compensated for by the increase of voltage gain at shorter wavelengths, thus making the overall gain of the antenna nearly constant over several octaves of frequency range.

A calibration setup for establishing actual gain figures for the antenna is shown in Trimpi's paper as well. This is in keeping with our assertion in Chapter 10 that actual meaningful antenna gain figures *can* be had! Trimpi's method can be modified and adapted to any receiving antenna, active or otherwise.

Trimpi's paper shows a fairly generic amplifier system that can be used with small loops, along with several example devices. Some of those particular op-amps may be difficult to acquire, but the OP-37 is common and readily available. I am particularly fond of the OP-37, probably because I've used more of them than any other op-amps. They work well at HF and are easy to work with. A more complete listing of suitable low noise RF op-amps, both current and obsolete, is shown in the Appendix, courtesy of Michael Trimpi. (Just because a part is no longer manufactured does not necessarily mean it is in short supply, or unavailable for experimentation. It just means you probably won't want to start mass producing any circuits that use them. For instance, I don't think the universe will ever run out of 1N34 germanium diodes, even though they haven't made any new ones in fifty years. There are probably still a few hundred billion of these devices in NOS — New Old Stock.)

If you are looking for a great general purpose HF receiving antenna, the Trimpi Loop is highly recommended. It is an antenna proven to perform in many scientific applications where the performance *has* to be verified.

The Resonance Fallacy

Like many other truths in Amateur Radio, the importance of resonance in antennas is often misapplied and misinterpreted. While many effective receiving antennas *are* resonant, sometimes even *very sharply* so, we need to be careful not to genuflect too deeply at that idol of resonance. There is absolutely nothing about a resonant antenna that causes it to radiate more efficiently than a non-resonant one. The "active ingredient" in antenna radiation effectiveness is *radiation resistance.*

As an example, *very few* AM broadcast towers are even remotely close to resonance at their operating frequency. Large towers come in 20-foot sections, typically. (Some older commercial towers came only in *60-foot* sections.) This means that if you're within 20 feet of resonance you're doing pretty well! Nobody prunes a broadcast tower to self-resonance. This means that a normal AM tower has considerable reactance at its feed point. Now, it *is* important to have *the system* as close to resonance as possible for the most effective power transfer from the transmitter. But this is taken care of by an efficient matching network at the base of the tower. Reactance has nothing to do with the actual radiating characteristics of the antenna.

Translating this concept to the receiving antenna, there is nothing about a self-resonant antenna that makes it receive any better. This is why aperiodic loops (as well as the "disappearing" whip) work as effectively as they do.

Null and Void

Like any small loop antenna, the Trimpi Loop has a deep null perpendicular to the plane of the loop. Depending on your particular needs, this may or may not be a problem. If you need an omnidirectional pattern, while maintaining all the other nice properties of the loop, the best solution is to erect *two* loops at right angles, and combine the outputs with a broadband 90° hybrid. That makes the pattern circularly polarized in the plane of orthogonality in the feed, linear in other planes.

The broadband 90° hybrid will be explored in more detail in later chapters in connection with the eXOgon antenna, but is useful for a wide variety of receiving applications. The MiniCircuits JSPQ-65W+ hybrid device is superbly suited to such applications. It operates from 5 to 65 MHz with very tight 90° phasing. It is only available in surface mount form, however, so you'll want the associated adapter that will give you access to SMA coaxial connectors. It is not a particularly inexpensive device, but is immensely useful for many receiving applications described in this book.

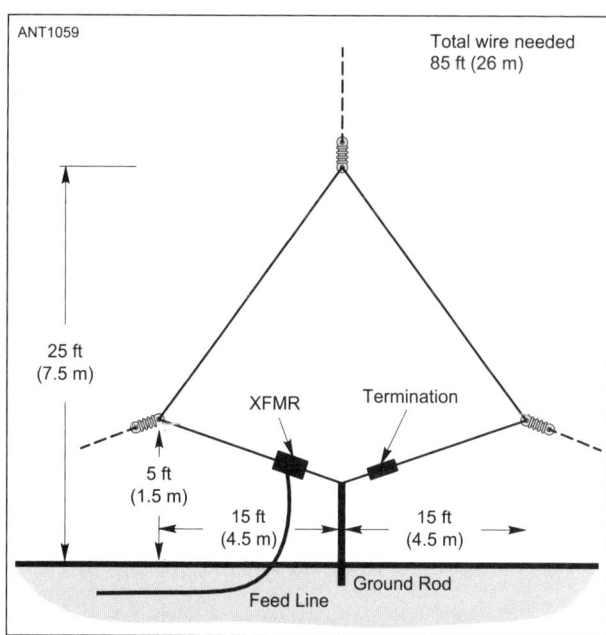

Figure 11.1 — Configuration of the K9AY loop at the maximum size that allows coverage of both the 160 and 80 meter ham bands with a resistive termination.

The K9AY Loop

A well known and popular aperiodic receiving loop in ham circles is the K9AY Loop (at least until multitudes of hams discover the Trimpi Loop!). As shown in **Figures 11.1** and **11.2**, the K9AY design is a terminated loop, normally configured in orthogonal pairs, to provide effective, symmetrical beam steering. Like the Trimpi loop, pairs of K9AY Loops can be combined with a 90° hybrid for omnidirectional operation, although this is not the normal means of deployment. The K9AY Loop is most useful for its impressive directional properties. The K9AY Loop can exhibit some impressive directionality for its size and can be erected with a single vertical support, which makes deployment fast and simple. This is especially convenient if you anticipated erecting *arrays* of

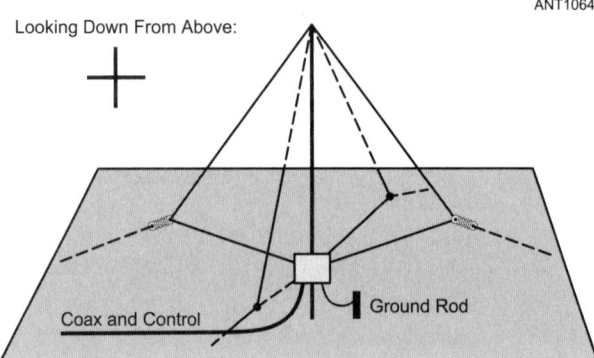

Figure 11.2 — Two K9AY loops can be installed with the same central support, creating a two-loop system that can be switched to cover four different directions. In a typical installation for 160 and 80 meter operation, the loops are 25 feet high and 15 feet from the center (30 feet across).

K9AY Loops, an arrangement with some very interesting possibilities.

Like any terminated loop, such as the pennant, flag, or EWE, the K9AY Loop has characteristics of both an antenna and a transmission line. The termination resistor balances out the transmission line behavior and the antenna behavior to achieve a null in the desired direction. While you can play around with the termination value to obtain some control over the pattern, most frequently the termination resistor is chosen to create the best null, and then (using switches or relays) to flip the resistor to the other "end" of the loop. With two cross-rigged switchable loops of this kind, four distinct directions around the azimuth can be chosen. Like most small antennas, the nulls are much sharper and more pronounced than the main lobes, so the K9AY Loop can best be described as a "steerable null" antenna.

Note that the Waller Flag, which usually operates at some height above ground, has been found to exhibit significant advantage in signal-to-noise ratio (SNR) in the horizontal configuration. Right at ground level, vertical polarization gives best results for both the K9AY and Waller Flag. Because of the way the K9AY loop is constructed, it is best deployed vertically on the ground. However, even as close as 7 meters to the ground, horizontal polarization gives improvements, so the Waller Flag configuration is preferred.

Not being one inclined to reinvent the wheel when I don't have to, especially since my wheel isn't likely to be much rounder than the other person's, I will present a nicely modeled *array* of K9AY Loops that someone else did. My favorite modeling program is *4nec2*, which is absolutely free for the downloading. It has a nice collection of antenna specimens in the database and one of them just happens to be a two-element K9AY Loop array for 160 meters. You can learn a lot by inspecting just one array like this. **Table 11.1** shows the *NEC* input file for the array. Now is as good a time as ever to review *NEC* modeling.

- The first set of rows, beginning with CM, are comments. This is pretty self-explanatory; here we have some of K9AY's original notes on the design.
- Next we have the GW lines, which are wire definitions. Each of these entries gives us the wire number, the number of segments in each wire, the beginning and end points of each wire (in X,Y,Z coordinates), and the wire diameter. (The entries for each row may be continued on the next line down. Each new entry will always start with a new GW prefix. All wire numbers in an antenna, or antenna array, have to be sequential. If your *NEC* program has a graphical entry interface, it will do the sequencing for you, but if you do it yourself, you have to be sure to get them all in order.
- The GS line is the units. *NEC* can specify meters, feet, or wavelength
- The GE tells us we're at the end of the geometry data.
- Next we have the LD rows, which specify the location and type of load and in this case we have 300 Ω resistors. The first entry is the type; a "1" here designates a parallel load. The second number is the tag, or wire number, the third entry is the *segment* on that wire number, which refers to the sequentially numbered wires. (This can be tricky to keep straight!) And finally we have the value of the load in RLC (ohms, henrys, and farads).
- Next we have the EX or excitation fields. This tells us where the generator(s) are connected, again in terms of wire number and segment. The next number, a 1 or a 0, specifies either a voltage or current source. And finally, we have the amplitude and phasing of each signal source.
- GN is the ground conditions. You can have no ground, perfect ground, or real ground, which is further specified by ground conductivity *and* dielectric constant. Ground conductivity is a complex number, something that's often overlooked. This topic is thoroughly covered in *Propagation and Radio Science*, written by yours truly and published by ARRL.
- And, finally, the FR row is the frequency, which can be specified as either a single frequency, or beginning and end points of a swept frequency.

Table 11.1
4nec2 Input File for K9AY Loop

See text for discussion.

```
CM The K9AY Receiving Loop (optimized for 160 m)
CM by Gary Breed, K9AY
CM first published in QST (ARRL), Newington, Sep 1997
CM The matching transformer is a 9:1 impedance,
CM 3:1 turns ratio type that should be familiar to many readers.
CM Five trifilar turns of ordinary hookup wire are wound
CM on a 3/4-inch diameter (0.825 inch) 43 material toroid.
CM The terminating resistor (Rterm) value will be between
CM 390 to 560 ohms depending on your band preference.
CM With average ground conductivity, a value of 390 ohms provides
CM the optimum F/B at 160 meters, while 560 it optimizes the loops
CM for 80 meters. A value of 470 W splits the difference for
CM "pretty good" performance on both bands. I chose to optimize
CM the antenna for 160 meter operation, so used a 390 ohms resistor.
CM Use an Rterm power rating of at least 1 W in case some transmitter
CM power ends up being coupled to the loops. I use a 2-W carbon
CM resistor, but two parallel 1/2 W or four 1/4 W units of
CM appropriate ohmic value can also be substituted.
CM Addition of another K9AY loop may significantly improve
CM the performance.
CM Converted with 4nec2 on 11-jan-03 by OK1RR
CE
GW 1 11 -0.070005 0 0.02434957 -0.0517428 0 6.08739e-5 6.24688e-6
GW 2 11 -0.0517428 0 6.08739e-5 -0.0334807 0 0.02434957 6.24688e-6
GW 3 11 -0.0334807 0 0.02434957 -0.0517428 0 0.03652435 6.24688e-6
GW 4 11 -0.0517428 0 0.03652435 -0.070005 0 0.02434957 6.24688e-6
GW 5 11 0.09131087 0 6.08739e-5 0.10957305 0 0.02434957 6.24688e-6
```

```
GW 6 11 0.09131087 0 6.08739e-5 0.0730487 0 0.02434957 6.24688e-6
GW 7 11 0.09131087 0 0.03652435 0.0730487 0 0.02434957 6.24688e-6
GW 8 11 0.09131087 0 0.03652435 0.10957305 0 0.02434957 6.24688e-6
GW 9 11 -0.0517428 0 6.08739e-5 -0.0182622 0 6.08739e-5 6.24688e-6
GW 10 11 -0.0517428 0 6.08739e-5 -0.0852235 0 6.08739e-5 6.24688e-6
GW 11 11 0.09131087 0 6.08739e-5 0.12783522 0 6.08739e-5 6.24688e-6
GW 12 11 0.09131087 0 6.08739e-5 0.09131087 -0.030437 6.08739e-5 6.24688e-6
GW 13 11 0.0913109 0.030437 6.0874e-5 0.0913109 0 6.0874e-5 6.2469e-6
GW 14 11 -0.051743 0 6.0874e-5 -0.073049 0.0243496 6.0874e-5 6.2469e-6
GW 15 11 -0.0517428 0 6.08739e-5 -0.030437 -0.0243496 6.08739e-5 6.24688e-6
GW 16 11 0.0913109 0 6.0874e-5 0.1156604 0.0243496 6.0874e-5 6.2469e-6
GW 17 11 0.0913109 0 6.0874e-5 0.1156604 -0.02435 6.0874e-5 6.2469e-6
GW 18 11 0.0913109 0 6.0874e-5 0.0669613 -0.02435 6.0874e-5 6.2469e-6
GW 19 11 0.0913109 0 6.0874e-5 0.0669613 0.0243496 6.0874e-5 6.2469e-6
GW 20 11 -0.051743 0 6.0874e-5 -0.073049 -0.02435 6.0874e-5 6.2469e-6
GW 21 11 -0.0517428 0 6.08739e-5 -0.030437 0.02434957 6.08739e-5 6.24688e-6
GW 22 11 -0.0517428 0 6.08739e-5 -0.0517428 -0.030437 6.08739e-5 6.24688e-6
GW 23 11 -0.051743 0 6.0874e-5 -0.051743 0.030437 6.0874e-5 6.2469e-6
GW 24 11 0.09131087 0 6.08739e-5 0.05783022 0 6.08739e-5 6.24688e-6
GS 0 0 164.27 ' All in WL.
GE 0
LD 1 5 1 1 300 0
LD 2 2 1 1 300 0
EX 0 6 1 0 1 0
EX 0 1 11 0 -0.866 -0.5
GN 2 0 0 0 13 0.005
FR 0 1 0 0 1.825 0
```

There are a few parameters we haven't included; the complete list of all *NEC2* parameters is included in the manual accompanying *4nec2*. Antenna modeling really does get easier with practice, and if you have a graphical input interface, it's even easier. Now that the antenna is put together mathematically, **Figure 11.3** shows what the thing looks like. And now we can plot the actual radiation pattern in 3D, as shown in **Figure 11.4**.

As you can see, there's a pretty nice null on the back of this array, and a pretty decent amount of forward gain too. This demonstrates the basic concept that if you want a lot of gain, you can either use one big antenna…or a lot of little ones. Arrays of antennas can be made of *any* basic antenna, and in any number, and sometimes it's just a lot more convenient to put a lot of little antennas out in the back forty than trying to construct one large one. But there are also things that you can do with arrays of antennas that you just can't do with larger antennas…such as very rapid beam steering. The phased array has replaced a lot of old school "big" antennas in fields as diverse as radar and radio-astronomy for just this reason.

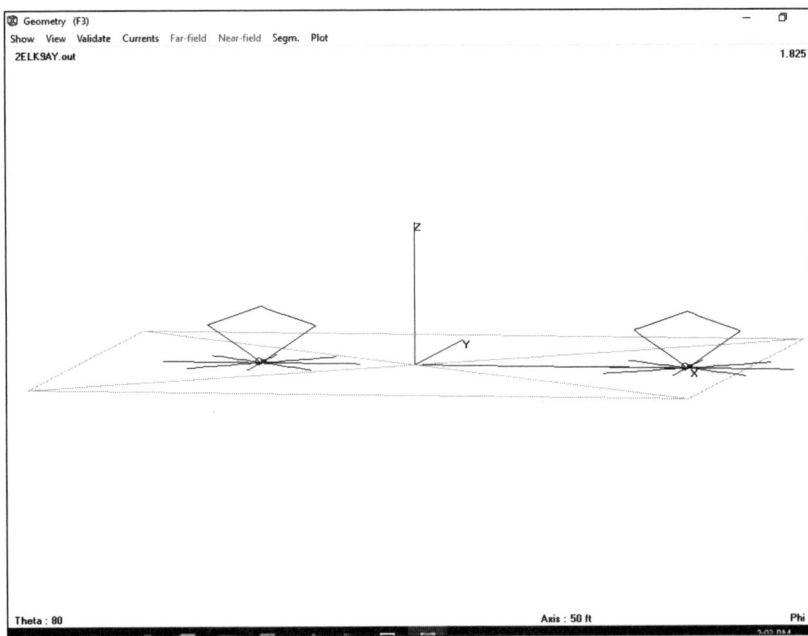

Figure 11.3 — A pair of K9AY Loops modeled in *4nec2*.

Figure 11.4 — 3D radiation pattern of a pair of K9AY Loops modeled in *4nec2*.

Narrow-Minded

One of the underlying themes of this book has been the broadbanded nature of the active antenna, and we certainly don't want to minimize the tremendous advantage the active antenna gives us in that regard. However, there are certainly times when you *do* want a narrow banded — or extremely narrow-banded antenna. As it turns out, the active antenna can shine in that application as well. And that is the topic of our next chapter.

Chapter 12

The Q Factor

One of the most important concepts to understand when it comes to antennas is this thing called "Q." Actually, Q applies to any number of ac circuits, lumped or otherwise, but it has a special application with regard to antennas.

Simply defined, for any *series* circuit, Q is the ratio of reactance to resistance, or X/R. Q is shorthand for "quality factor." One needs to be careful when flinging around terms like "quality" however. Depending on what you're trying to accomplish, a circuit with a high quality factor may be the last thing you want. In its simplest incarnation, Q is an indicator of a circuit's frequency *selectivity*.

Now, before we go too far down this rabbit hole, we should probably bring up a few intriguing points. We know that, by definition, a *resonant* circuit is one in which the inductive reactance (X_L) is equal to the capacitive reactance (X_C). In such an instance, the circuit can be replaced by a simple resistance, or more precisely, its *impedance* is equal to the resistance alone.

Since the net reactance of a resonant circuit is zero, it might be tempting to conclude that the Q of any resonant circuit, at resonance, is also zero. This is obviously not the case. Or, it *should* be obvious, at least to "properly seasoned" radio amateurs. To mitigate this possible confusion, however, it's always been "safest" to define the Q of any series circuit as being X_L/R. This way, we may retain the "true Q" of any circuit. Also, from a practical standpoint, it is the quality of the *inductor* that usually determines the Q of the circuit. All practical inductors have some measureable (and often significant) series resistance in them, while the parasitic resistance of most capacitors used in radio circuitry is insignificant by comparison.

There are times when you want a very narrow-banded antenna. But first we should probably define what we mean by narrow-banded. Normally, when we think of the bandwidth of an antenna, we think about the frequency range over which the antenna is more or less resonant. Although not the best indicator of resonance under all conditions, the SWR on the transmission line (assuming you actually are using a transmission line) is a fair indicator of resonance. Generally speaking, the farther you move *away* from resonance (specifically when transmitting), the greater will be the SWR on the transmission line (assuming that at resonance the antenna feed point resistance matches the feed line impedance). So, we are conditioned to think of the bandwidth of an antenna as the frequency range over which some SWR value is not exceeded. We may speak of a 2:1 SWR bandwidth, or a 3:1 SWR bandwidth, for instance. What value of SWR is deemed acceptable is pretty much arbitrary. But since it's fairly easy to measure, most of us are "comfortable" with this kind of antenna evaluation.

This is fairly useful for a simple antenna such as a dipole. Where it gets a little stretched is when we apply it to something like a high gain Yagi, where there can be a great difference between *pattern bandwidth* and SWR bandwidth. This has important practical applications: The antenna has a pattern/gain bandwidth, and impedance bandwidth, and an efficiency bandwidth, and the three might not coincide. Pattern bandwidth is the frequency range over which an antenna maintains some prescribed pattern or gain. Some "high-strung" Yagi designs can have extremely narrow pattern bandwidths, wherein the gain or directivity deteriorates very rapidly as you move off the design frequency. In reality, the "bandwidth" of an antenna is the overlapping of these two parameters, SWR bandwidth and pattern bandwidth. When you overlap these two requirements, you may find that your antenna's "usable" bandwidth is far less than either the pattern bandwidth or SWR bandwidth alone.

With all that being said, having an extremely narrow antenna bandwidth (of either sort) is not necessarily a bad thing. We'll talk about when having an extremely narrow-banded antenna can be a great thing…and how active antenna principles can make it even greater.

It's No Contest

It's fascinating to observe from afar, the development of Amateur Radio receivers over the past century. (This is one of the best reasons to be an ARRL member, as it gives you online access to all the *QST* articles back to 1915.) Receivers evolved from simple coherer detectors, which would pretty much copy anything and everything that remotely resembled a radio wave — simultaneously — to a progression of increasingly more selective receivers. The regenerative receiver was capable of extreme selectivity, as well as sensitivity — but it took the touch of a safecracker to operate. The superheterodyne radio came to the rescue, though in chunks, as progressively more conversion stages were added. These developments weren't without their drawbacks, however.

In relatively recent radio history, this concept known as "dynamic range" came into view. As more and more hams attempted to break DX pileups or win contests (or, found themselves being the actual *object* of a pileup), the ability to separate signals was understood to be more than a simple matter of…well…*selectivity*. The ability of a receiver to not overload in the presence of strong signals, while trying to decipher a *weak* signal, became an unobtainable goal. Or, at least it seemed to.

Without going into too much gory detail about receiver specifications, dynamic range, intercept points, and other more arcane topics, all of which are covered fully in the *ARRL Handbook* and elsewhere, it turns out the "answer" to an excellent receiver is fairly simple. You want to keep the selectivity as close to the antenna as possible. This discovery was at odds with a long progression or receiver progress, the trend of which was to have progressively higher levels of selectivity at progressively lower IF frequencies. The direct conversion receiver bucked the trend, with some astonishing results.

All active devices have some intermodulation distortion (IMD). The more signals an active device has to contend with at the same time, the more intermodulation products are possible, and thus *dynamic range* suffers. The goal is to reduce the demands of your receiver's active devices as much as humanly possible, which means filtering out as many signals as you can with a passive device, which has no intermodulation distortion.

An antenna is a passive device — it can produce no intermodulation distortion on its own. (This is not *strictly* true, as cell phone companies have recently discovered this obnoxious phenomenon called *passive intermodulation distortion* or PID, caused by dissimilar metals used in various RF connectors. However, this particular phenomenon is essentially absent in HF receiving applications.) The bottom line is, if you can possibly filter out unwanted signals with a linear (electrically) piece of wire, it's a lot better than trying to do it with a nonlinear transistor — or a lot of nonlinear transistors. Or to restate the case, if you can use antenna Q as your first line of defense, everything will be happier.

Some Defining Moments

One of the nice things about a loop antenna, especially a *small* one, is that it behaves much more like a normal electrical circuit than, say, a dipole. "Lumped constants" are just a lot easier to analyze than distributed capacitance and inductance. And, methods for manipulating these lumped constants are much easier and effective in most cases.

To review, the Q of a circuit is defined simply the ratio of reactance to *resistance* (or better yet, stored to dissipated energy). I say "simply" because, while the math is simple, the implications are profound.

So we don't wander too far, let us first describe inductance in terms of a simple *solenoid* winding, uniformly wound, with just one layer. For any simple solenoid, the inductance goes up as the *square* of the number of turns. For the same simple solenoid, the resistance of the total wire goes up in direct proportion to the number of turns. If you double the turns, you have twice the resistance, but four times the inductance. But the inductive *reactance* is in itself a function of frequency. This means for any *given* coil, the Q is going to go up proportionally with the frequency. Well, that works up to a point. As the frequency is increased, the *skin effect*, which is a resistance, also increases. What this means is that at some point the increase in skin effect due to increasing frequency cancels out any increase in Q you might achieve by cranking up the frequency. Exactly *where* this point occurs is a very complex matter, depending on all kinds of variables like wire gauge and winding pitch. We don't need to go too deep here.

What we *do* need to know is that, for any given *frequency*, the Q of a coil will increase with the increase in inductance. The bottom line is: TCC or Total Copper Content. The more turns of wire you have on your solenoid, the greater the Q.

If you're an astute student of radio, about now you will be asking "what about capacitance?" Glad you asked, because Q without capacitance doesn't do much for us in radio. Half a resonant circuit is…well…half a resonant circuit. So, let us define the resonant frequency as the frequency at which the inductive reactance and capacitive reactance are the same. When this condition prevails, the reactances cancel, and then, all that's left impedance-wise in the circuit is the resistance. And, in the case of a really small loop antenna, it's all *ohmic* resistance; any radiation resistance is nonexistent.

Parallel Universe

When talking about resonant loop antennas, someone will always ask, "Is that parallel resonance or series resonance?" That's a good question. If you only have two components, an inductor and a capacitor, and they're wired together in the only possible way two components can be wired together, are they in series or in parallel?

Resonance is resonance. You are encouraged to ruminate on the "The Q of Everything," included in the Appendix at the end of this book. But, what about that "other formula" for parallel resonance where Q is resistance divided by reactance (R/X)?

Ahh. But then I have to ask, "Which resistance are you talking about?" If you're referring to an *external* resistance, that is, a resistance *across* either the C or the L, that has some meaning. But we aren't talking about *loaded Q*. In the case of our small loop, the only resistance is *internal*, and so the "normal" Q formula strictly holds…at least until we try to suck some power out of it.

By the way, while the loaded Q issue is not directly related to receiving antennas, there is enough confusion about the matter to merit a full treatment of the topic in the Appendix, in "Get a Load of This: Taking the Mystery Out of Loaded and Unloaded Q." We trust that piece will be interesting reading.

Noise That Annoys

This matter of Q has a special significance in regard to receiving antennas. While antennas with very high Q are generally avoided in transmitting applications (with the notable exception of small transmitting loops), very high Q antennas can be put to great advantage in many receiving situations.

High Q antennas (in any form, not just loops) tend to be lower noise than equivalent antennas with lower Q. Noise, by its nature, regardless of its source, falls into a broadband character. Whether manifesting itself as a group of well-defined RF harmonics, as in the case of power line noise, or noise emitted from some types of switching regulators, or in the form of RF "white noise" (which is more evenly distributed), it can almost always be reduced to some degree by means of a high Q, selective antenna. Not only are signals outside of the desired frequency "filtered" by means of the antenna's selectivity, but the actual *gain* of the antenna can be significantly increased by increasing the Q. A high Q tank circuit, which includes any tuned loop antenna, accumulates energy over a large number of RF cycles.

Larger inductance/capacitance (L/C) ratios generally result in higher Q circuits, all other things being equal. We know that the inductance of any coil increases as the square of the number of turns, while the "ohmic" resistance only increases linearly with the total length of wire. This tells us that for any given wire resistivity, we will have a higher Q circuit with more inductance. For this reason, we find that *multi-turn* loops have higher gain than loops of equivalent dimensions. While the "true" aperture of a loop antenna is determined by the "amount of sky" it occupies, multiple turns around the same perimeter result in higher induced voltage at the antenna terminals, which is almost always advantageous, especially when feeding an infinite impedance amplifier. Extending this concept even further, we find that the many-turn ferrite rod "loopstick" antenna can have a much greater *effective aperture* than one might expect.

So, now that we've established that resonance is resonance and Q is Q, how do we best apply this knowledge to receiving antennas?

The higher the Q of a circuit (antenna or otherwise) the more sharply resonant it is. This has two effects. First, the antenna is more selective. Signals outside the resonant frequency of the antenna drop off more quickly as you move away from dead center. Second, there is an increase in terminal voltage. A high Q antenna exhibits voltage gain, and it can be substantial. In fact, for a resonant circuit, the voltage step-up is precisely equal to the Q. Assuming we feed said high Q antenna into an infinite impedance load, we can expect considerable signal gain over a lower Q antenna.

Does this mean we get something for nothing, such as power gain from a passive piece of wire? Well, no. What a high Q circuit does is accumulate energy from successive cycles of the exciting wave. We've taken a lot of little RF "pulses" as it were, and allowed them to stack up on top of each other. It's the same principle in a swinging pendulum. You can give the pendulum tiny little taps during each swing, and the amplitude of the swing will keep increasing. The overall effect isn't so much that a high Q antenna is "bigger," it's that it's an antenna that's in operation *longer.* We can store energy from more successive wave peaks than we can with a lower Q antenna. Again, the law of NFL (no free lunch) strictly applies.

Proper Proportions

Using what we know about solenoids, it would stand to reason that the more turns of wire you have, for a given size coil form, the more inductance, and hence, more Q you will have. And this is indeed the case. High-Q multi-turn small loops such as the one shown in **Figure 12.1** consist of many turns of wire, and a relatively small amount of capacitance. And, again, as long as the loop is very small, relative to a wavelength, the radiation resistance will be negligible, so the only resistance we need to deal with is the wire "ohmic" resistance. A very different case from a transmitting antenna, where we actually want to *maximize* radiation resistance!

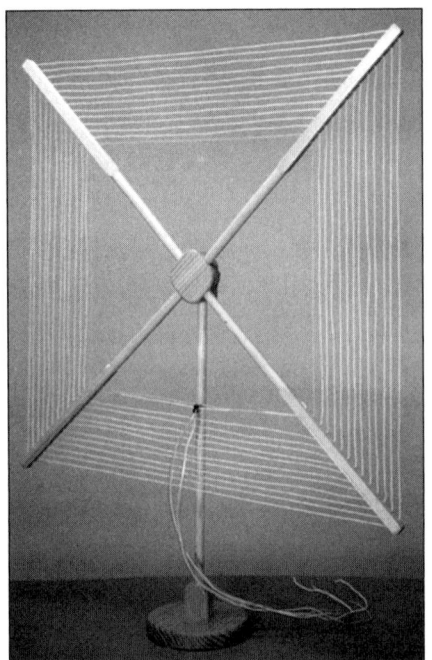

Figure 12.1 — A small multi-turn loop receiving antenna. [Courtesy Peter Jennings, AB6WM]

The Golden Ratio

Back before we had much in the way of computers other than the slide rule, a gentleman by the name of Wheeler worked out the optimum dimensions of a solenoid for maximum Q. The so-called Wheeler inductor having this Golden Ratio is one that has the most possible inductance for a given length of wire. The Wheeler coil is a little bit shorter than the diameter of the coil, on the order of 90%. This is actually a rather short, fat coil. The Wheeler formula has been slightly revised with the advent of numerical computer modeling, but he got it pretty close.

Now, while a receiving antenna built according to the Wheeler formula might be extremely selective, it may not be very *sensitive*. If you recall from the previous chapter, the *gain* of a loop is proportional to the enclosed area of the loop. So we have a little bit of a "conflict of interest" when designing a high-Q coil that also needs to serve as an antenna. Most practical multi-turn resonant loops, such as the one shown in Figure 12.1 are much fatter and shorter, proportionally, than the Wheeler inductor. They resemble a pancake more than a cylinder.

The Core of the Matter

While the multi-turn resonant "small loop" can indeed be very small

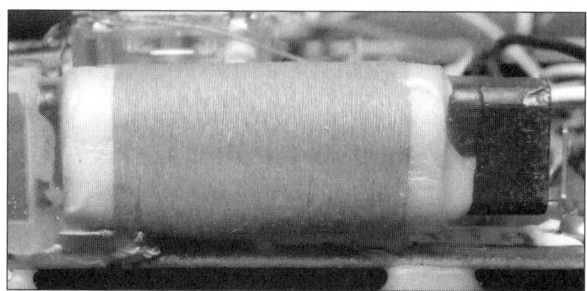

Figure 12.2 — Typical ferrite core antenna for the AM broadcast band as found in an AM/FM portable radio.

and effective, it can be made even more so using a method that is utterly unavailable to the transmitting antenna, namely the high-permeability *magnetic core*.

So far, all our discussion about loop antennas has assumed an *air core* coil of some kind. At lower HF frequencies and below, it becomes very practical to wind a receiving antenna around a ferrite core, usually in the form of a rod. Such an antenna is known as a *loopstick* antenna, and has some rather interesting properties. A typical loopstick is shown in **Figure 12.2**.

First of all, the core greatly increases the *permeability* of the coil, allowing a much larger amount of inductance for a given length of wire. While utterly unsuitable for transmitting, the loopstick antenna is capable of astonishing performance on the lower amateur bands.

The loopstick antenna has been almost universal since the advent of the transistor AM radio, but it has been greatly underutilized in the Amateur Radio sphere. Not only is it capable of great selectivity and sensitivity in a minuscule package, but it also excels when it comes to direction-finding and nulling, having a much cleaner pattern than its larger counterparts.

When using a multi-turn loop antenna of either the free air or loopstick variety, it is important not to spoil the inherent Q of the loop when you attempt to couple it to a receiver. There are at least a couple of means of achieving this end. The important thing is to *not* try to achieve maximum power transfer between the resonant loop and the transmission line (if used) and the receiver. The traditional method is to use a secondary *loose-coupled* loop, usually of no more than one turn and often much smaller in diameter than the main loop. Again, the coupling needs to be *much* less than that necessary for maximum power transfer.

A second method is to connect the output terminals of the loop to a near-infinite input impedance amplifier. The AD8067 is ideally suited to this task, just as it is to the active whip. The advantage of the active method is that you won't suffer the loss of signal experienced when using very loose inductive coupling. If you really want to obsess about the matter, you can feed the loop terminals to a balanced pair of AD8067s in order to maintain absolute symmetry. This is highly recommended when using a loopstick for direction finding. Just such a project is described in detail later in the book.

The directional properties of the loopstick are exactly the same as for an air-core loop — maximum sensitivity edgewise to the loop, and a perfect null perpendicular to the loop. However, in a loopstick antenna, the *core* is perpendicular to the loop. So to null a signal, you want to aim the *core* in the direction of the incoming signal. A longer core has a better defined null than a shorter one, and if possible, should be a few times longer than the winding itself.

Snake Oil

Despite the remarkable performance of the loopstick antenna, since the advent of the transistor radio, a number of magical mystery properties of the loopstick have been attributed to the device, completely contrary to the laws of physics. Advertising terms such as "Wave Magnet" have contributed greatly to the misconception.

There is no known device in the universe that can "pull in" or "attract" radio waves. Radio waves go where they go, and if you happen to be in the line of fire, you can intercept them. If not, they just pass right on by. Yes, a highly permeable core will concentrate the flux lines *inside* the core, and yes, this has tremendous advantage in enhancing the Q and sensitivity of the device. But no, the core can have no effect on radio waves merely passing through the neighborhood.

This fact probably has no practical effect on the actual practical use of the loopstick, but it's a good idea to get the science right, too!

I Want Arrays

Because of the extreme compactness of the ferrite loopstick, it is immensely convenient to build large *arrays* of the devices. This is a nice thing to have when building instruments such as interferometers, especially for doing vertical angle of arrival measurement. You can distribute a number of ferrite loops up a tower and measure the relative phase angles of the signals arriving from them, and thus determine the vertical angle of arrival. You can also conveniently deploy a number of *space diversity* schemes, which would be highly unwieldy with full sized antennas — especially on the lower frequencies.

Closing the Loop

We continually get comments from hams, even "well-seasoned" ones, who are astonished at the performance of the small loop receiving antenna — active or passive. Unless you actually try one out, it's easy to discount its performance, and it solves a number of issues that would be difficult to resolve any other way. If you have a noise problem that has kept you away from the lower bands, a small loop antenna may be the only answer you need. It's definitely worth a try.

Chapter 13

The Beverage: In a Class of Its Own

It is fitting that we follow our chapter on tiny antennas with a chapter on huge antennas. It serves to demonstrate that there is something for everyone when it comes to receiving antennas. Not everyone has the acreage to put up something like a low-band Beverage antenna…but again, some of us do.

Before we go too far into this, we need to give credit where credit is due. Dr. Harold Beverage was one of the pioneers of RCA, developing methods for commercial transoceanic communications. He had a grasp of what a receiving antenna needed to do long before just about anyone else. He had erected a vast array of his groundbreaking *wave* antennas at RCA's Riverhead, New York, research facility. His original work was done at frequencies far below our 160 meter amateur band, but the principles have been successfully adapted to "our" use for decades.

A Beverage antenna (**Figure 13.1**) consists of a wire, typically 1 wavelength or longer at the lowest operating frequency — about 550 feet for the 160-meter band! Longer antennas provide increased gain and directivity, but shorter antennas can be useful as we will see later in this chapter. Beverage antennas are installed at relatively low heights, normally 8 to 10 feet above ground. They should form a relatively straight line extending from the feed point toward the preferred direction. There is

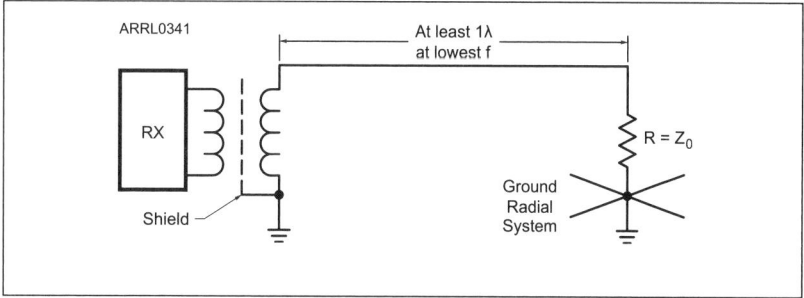

Figure 13.1 — The one-wire Beverage antenna forms a transmission line with the ground and is terminated at the far end. The antenna's preferred direction is to the right in this drawing.

a terminating resistor to ground at the end farthest from the radio. The terminating resistor absorbs the power of signals arriving from the unwanted direction instead of allowing them to be reflected back toward the feed point. Beverage antennas are effective directional antennas for 160 meters and 80 meters and our newest bands at 630 and 2200 meters. They are less effective at higher frequencies, however, and are seldom used on 40 meters and shorter wavelength bands.

How It Works

The Beverage is a *wave antenna,* with several fairly unique and possibly unfamiliar properties. The directionality is completely different from a "normal" antenna. If you were to suspend a straight horizontal wire in free space, it would have absolutely no response directly off the ends. A wave traveling "head on" to the wire has both E field and H field perpendicular to the wire. No component of the wave can induce a current *in the direction* of the wire.

Now, let's take the same wire and suspend it over a parcel of lossy earth. If the earth were perfectly conductive, the wire would simply form half of a "leaky" transmission line, the earth below it forming the other half. Likewise, it would be unresponsive to signals directly off the end… or from any other direction, for that matter. But if the earth below is a dielectric, instead of a perfect conductor, radio waves traveling along it will be slightly slower than a radio wave in free space. But even more interestingly, a vertically polarized radio wave on a grazing path with the earth will tilt slightly in the direction of travel, because if the top part of the wave is moving faster than bottom part, it *must* bend in the direction of the slower part — or the wave will *break,* a physical impossibility. Because the wave is now tilted forward slightly, it now has a very small

---

Wave Antennas

One type of antenna that clearly delineates the differences between transmitting and receiving antennas is the *wave antenna* in its various incarnations. In fact, there is really no analogy for a wave antenna in the transmitting realm. A wave antenna works by adding up accumulated *phase shifts* of induced currents as the wave passes through the antenna. Wave antennas such as the gargantuan Beverage make *lousy* transmitting antennas, but have excellent low-noise receiving characteristics. Likewise, the signals extracted from wave antennas are typically very small, and can benefit from active antenna methods.

electric component in the direction of travel. If it approaches the wire end on, this minuscule vector in the direction of the wire will induce a small voltage. As it turns out, the voltage induced is proportional to the length of the wire. The longer the wire, the more the signal, indefinitely.

As you can see, the Beverage is thus very strongly vertically polarized. Therefore it strongly rejects any signals *not* arriving end-on.

Also, as you can probably imagine, the Beverage antenna is not a high gain antenna, but it is a highly directive antenna in the preferred direction, which is a far more important property, at least as far as receiving is concerned. In addition, it is a very *quiet* antenna; something Dr. Beverage understood as being the key to low-frequency communications. The Beverage can benefit, therefore, from a good low-noise amplifier right at the feed point. The Beverage deployed as an *active antenna* can be a formidable competitor.

Because the Beverage is normally erected at a height just sufficient to clear moose antlers (at least in Alaska), it functions partially as a transmission line as well, and should be terminated at the far end to prevent standing waves from modifying its normal unidirectional pattern.

Like any other antenna, the Beverage may be part of an array, as well. I don't know of too many hams that have the real estate for *arrays of Beverages*, but for those that do, that possibility exists, as well. Barring that, any Beverage can be steered in at least two directions by swapping the termination resistor and the feed point. The *ARRL Antenna Book* shows a fairly simple arrangement for doing this.

Getting Bogged Down

The Beverage on Ground (BOG) is a variation on the Beverage which can be surprisingly effective…or not. For some interesting reading, please refer to the July/August 2016 *QEX* article by Rudy Severns, N6LF. Rudy is a key player in the ARRL 600 Meter Experimental Group (**500kc.com**), has experimented extensively with low-band antennas, and knows whereof he speaks.

As the name suggests, the BOG is a Beverage made by laying a wire right on the ground. As you might suspect, the wire has to be insulated, and it is somewhat more dependent on local ground conditions than the normal Beverage. A typical BOG is around 200 feet long and can sometimes be a viable alternative to a traditional Beverage when logistics or available space prevent the latter.

In interior Alaska, the ground conductivity is *so* bad that it's like living in free space. An 80 meter dipole works just fine lying on the ground — for both transmitting and receiving. Likewise, a BOG tends works up here a lot better than it works other places. A BOG is *almost always* operated as an active antenna, however, since the signal levels are much lower than even a normal Beverage. In any case, recognize that a BOG antenna is very much YMMV (Your Mileage May Vary).

Modeling the Beverage

The Beverage is one antenna that particularly yields itself to modeling, if you happen to have access to *NEC4*, which supports lossy ground connections. However, before you can model your dream Beverage over ground, you have to know what your ground conditions actually are — and that can sometimes be tricky. For decades, the FCC has published charts of ground conductivities and dielectric constants for the continental US. This was originally done very early in the AM broadcast game, and for decades, hams in the know have relied on these charts. However, what they fail to show is how rapidly ground conductivity can change over surprisingly small distances. The original FCC charts were eventually found to be nowhere near "fine-grained" enough to be of much practical use for amateurs. (Not to mention that this original work was done before Alaska was part of the US, so we KL7s were conveniently omitted. Not that it would have helped much, because our average ground conductivity here is about four orders of magnitude *worse* than the poorest soils in the continental US. Most of the time if we just enter "very bad" in the ground parameter of the modeling software, the results are pretty accurate.)

A very interesting distributed science project would be for countless hams all across the country (and around the world) to perform accurate local ground measurements. It would be a great way for us to contribute to the "advancement of the radio art," which is one of the things we're supposed to do as hams. A little food for thought, perhaps.

Tuning

While the Beverage is inherently a very broadband antenna, not having any resonances or standing waves (in theory), the pattern can be largely dependent on frequency. We know that it is progressively more directive as the frequency is increased. Various flavors of the Beverage, such as the multiwire Beverage, were developed to flatten out the performance with respect to frequency, as well as to perform a little bit of beam steering as in the *echelon* array.

Because of the broadband nature of the Beverage, there is no crucial tuning involved; it will almost always work, and work well. It lends itself to easy experimentation; all you need is some wire and a lot of real estate. (Actually the real estate requirements have always been exaggerated a bit;

as long as the Beverage is more than ¼ wave long at the lowest frequency of operation, it *will* operate as a Beverage. It just gets *better* as it gets longer.)

Feeding the Thing

I'm not normally much for *buying* anything that resembles a wire antenna; I believe knowing how to string wires up and make them sing to the ether is a fundamental skill for every ham. However, if you *must* pay good money for a piece of wire or two, at least get something innovative. DX Engineering offers a reversible Beverage kit, DXE-RBSA-1P that includes a reflection transformer, termination box and a generous length of 450 Ω window line. This is a two-wire Beverage set, which when properly installed is capable of bi-directional steering with very deep nulls. You can set up *two* of these antennas orthogonally for four-direction beam steering…if you have the space.

Designing the reflection transformers for two-wire Beverages can be a bit tricky, so the pre-fabricated transformers included in these kits are an advantage. You can be forgiven for violating my no-buy rule when it comes to wires.

A somewhat less "tricky" method for building a bidirectional Beverage is to simply run two transmission lines out: one to the near end and one to the far end. However, you'll also have to rig up some way of switching the termination to the opposite end of the antenna. The inconvenience of this is what basically instigated the two-wire method which actually was borrowed from ancient telephone company "phantom circuit" topology.

Either method works fine; which you use will depend on the amount of coax you want to drag out to the back forty.

Short Beverages

The Beverage antenna deserves a chapter of its own, because it is quite a unique antenna, even for a "wave" antenna. There's still a lot of "unfinished business" when it comes to fully optimizing the Beverage antenna. While Beverages typically use up a lot of real estate, "reasonably" sized Beverages can be used to great effect with modern low-noise active electronics. The continuously loaded or "Slinky" Beverage produces more wave tilt per foot than a straight piece of wire (though in the opposite direction), and can be made a lot shorter for a given output signal, while still maintaining its unique "Beverage" properties. Such a hybrid Beverage merits a lot more experimentation from innovative hams.

Some Terminal Thoughts

Because of the variability of ground conditions, the optimum value of termination resistance will have to begin with an educated guess. On the average, a value of 800 Ω will give you a functioning Beverage, but you will have to experiment with that value to optimize the front-to-back ratio of your particular Beverage. Most likely, once you have one Beverage optimized, the value should be right for any and all other Beverages on your property, but don't count on it. As we suggested earlier, ground conditions can change dramatically in just a few hundred feet. This is a good field for experimentation.

If you'd like to learn more about the classic Beverage antenna from its creator, check out the following resources:

- **www.hard-core-dx.com/nordicdx/antenna/wire/beverage/interview1.html**
- H.H. Beverage, "The Wave antenna for 200 Meter Reception," *QST*, Nov. 1922, pp. 7 – 15.
- H.H. Beverage and D. DeMaw, "The Classic Beverage Antenna, Revisited," *QST*, Jan. 1982, pp. 11 – 17.

Chapter 14

Dealing with Non-Reciprocal Propagation

"Everybody talks about radio propagation, but nobody does anything about it." That is a paraphrase of a more famous commentary on the weather. But it's true. Radio propagation is one of the most popular topics of discussion among radio amateurs, and since it's so central to most of what we do as hams, it's somewhat ironic that we have made so little progress, technologically in this area.

This chapter will be more about operating methods than about technology. We may not be able to "fix" radio propagation directly, but we can understand a few things about it that will allow us to optimize the use of the physics we're dealt with.

Hopefully by now, we've been able to effectively demonstrate that transmitting and receiving antennas can be different, sometimes profoundly. As we have discussed, the concept of *reciprocity* is firmly entrenched in the physics of electromagnetism, right down to the molecular and atomic level. In free space, reciprocity is absolute, as well as an extremely useful property. Reciprocity is what allows us to model receiving antennas with the same algorithms used for transmitting antennas. But what we still haven't touched on yet is that radio propagation paths themselves can be very different, depending on which direction the radio waves following them are traveling. Here again, we want to use the term bilateral when speaking of complex propagation paths, to distinguish it from the more fundamental *reciprocity*. Radio antennas may be *reciprocal* in a number of ways, but most real world propagation paths aren't *bilateral* — at least where the ionosphere is involved.

If HF radio propagation were truly reciprocal, we could eliminate *one* reason for using separate transmitting and receiving antennas. However, when we understand that non-bilateral propagation is the *norm* for HF radio, we will understand that we can benefit by using separate transmit and receive antennas even if they're *exactly the same type*.

Your Ground is Not My Ground

There are countless factors that can result in non-reciprocal HF propagation. Not only is the ionosphere often non-reciprocal (or perhaps, another useful term might be "non-bilateral"), but ground conditions can also greatly affect the reciprocity of radio propagation.

The Earth is not a homogeneous or highly conductive surface. It would certainly be nice if the Earth's surface was made of copper, but it's not. Radio waves at HF and MF frequencies can penetrate the earth to an astonishing degree, and the character of the ground far beneath the surface can have a profound effect on overall radio propagation.

Ground conductivity, both local and distant can and will affect radio propagation. In addition, abrupt changes in ground conductivity along a radio path (such as the interface between the seashore and the sea) can result in highly non-reciprocal ground propagation. (In this case, we are including sea-surface propagation in the whole ground wave discussion.) Just as in the case of skywave reflections over a long path ("Your sky is not my sky"), we cannot assume reciprocal ground conditions along a long radio path…or even a relatively short one.

All this contributes to the interest (and sometimes frustration!) of HF radio propagation. There is still much to learn.

The Bent Sky

The ionosphere has lots of wrinkles. This is not too surprising when you realize that it's primarily made up of *gas*. It is *not* a hard target with a well-defined reflecting surface. The fact that we can get *any* kind of coherent reflection, *ever*, from such a transient and fickle medium is amazing in itself. In fact, the term *reflection* itself is seldom an accurate model of what's actually happening. While the end result of a refracted signal may appear to be a reflection from a purely *geometric* perspective, it tells us very little of the internal mechanism of the ionosphere.

Now, as warped and wrinkly as the ionosphere may be, one still might possibly and stubbornly assume that any propagation path between A and B would be reciprocal. It may be a circuitous path from A to B, but it, one might argue, should be the *same* circuitous back from point B back to A.

If there were no magnetic field, this assumption would be plausible. However the Earth's magnetic field changes everything. And it changes everything a *lot*.

A magnetized plasma (which describes the Earth's ionosphere), first and foremost, has two different refractive indices. A refractive index is the velocity of electromagnetic radiation relative to the speed of light in free space. A non-magnetized plasma will slow radio waves down relative to the speed in free space. And the *density* of that plasma will determine just *how much* that radio wave will be slowed down compared to free space. In the non-magnetized state, the refractive index *is* indeed reciprocal, and you can use some fairly intuitive optics to determine the paths of rays through this medium.

On the other hand, a plasma in the presence of a magnetic field has *two* different refractive indices, depending upon whether the radio wave is running *parallel* to or *perpendicular* to the magnetic field lines. This does two interesting things. The first is that it causes a linearly polarized radio wave to separate into two counter-rotating circularly polarized waves. One of these waves, called the O-mode (for ordinary) is right hand circularly polarized in the Northern magnetic hemisphere. A complementary ray is also generated, known as the X-mode wave (for extraordinary), which is left hand circularly polarized in the Northern magnetic hemisphere.

These two circularly polarized rays diverge away from each other and from the magnetic field lines. This produces two obvious effects of interest to radio amateurs. The first effect is that neither one of these rays cares one iota about great circle paths!

Not only can magnetic fields cause birefringence, but they have the more obvious effect of bending, tilting, and warping the ionosphere. Even neglecting, for a moment, the non-reciprocal properties, the fact that the ionosphere is neither even nor horizontal can cause reflected or refracted waves to be skewed in azimuth, as well as elevation. When you add the birefringence to the mix, HF propagation can become extremely complex and convoluted.

Model Behavior

Of all the readily available propagation prediction software out there, only one program takes into account the Earth's magnetic field: *Proplab Pro* (**www.spacew.com/www/proplab.html**). *Proplab Pro* is a true *ray tracing* program that actually shows the independent paths of the X and O waves between any two points on Earth, using real world conditions. This program is described in greater detail in my book *Propagation and Radio Science*. Several examples in that book show just how radically the X and O rays can diverge from one another, especially near the Earth's magnetic poles. Near the magnetic equator, the divergence is a lot more moderate.

So What

Our intent in this chapter is not to cover the physics of ionospheric propagation, which has been covered well elsewhere. Rather, we will explain what you can do about it using the hardware you have available. The first thing we need to do is to acknowledge that HF propagation paths are *not* the same in both directions, and adjust our operating habits accordingly.

Since the advent of directional antennas, standard practice has been to "home in" on a DX station by steering your beam for the best received signal strength, with the assumption that this is also the best direction to be transmitting. In some rare cases it may be, but in the vast majority of DX communications, this is not the optimum state. The only way to confirm this, of course, is to ask the DX station for some feedback as you rotate your transmitting antenna. Of course, if you're using the same antenna for transmitting and receiving, this could present some logistics issues; at the very least it might cause some undue wear and tear on your antenna rotator as you swing your antenna back and forth for best results between every exchange.

The obvious, better, solution is to have a separate receiving antenna that can be optimized for the best incoming signal without disturbing the best orientation for your transmitting antenna.

Because the vast majority of radio amateurs that *do* have highly directional antennas use the same antenna for transmitting and receiving, one can be forgiven for never encountering the azimuth skewing phenomenon. It's probably safe to say that 90% of even experienced operators rely entirely on great circle paths to set their antenna azimuths. If you aren't looking for it, you could miss it…throughout an entire long ham "career." (I confess I was entirely oblivious to this effect until I started working at HIPAS Observatory and HAARP. This X and O business has been known by radio scientists for most of a century; it's just taken a while for hams to finally catch on.)

And

If things weren't complicated enough, here's another point to consider. We've already established that every transmitted HF signal will be split into two divergent rays, one being an O wave and the other being an X wave. However, from the receiving end, you really don't know which of these you're actually dealing with. You can possibly make an educated guess. But to really know, you need a *circularly polarized* receiving antenna to discriminate between the two different modes. How many hams do you know who are using circular polarization (CPOL) on HF? We aim to change these pathetic statistics with our unique eXOgon antenna, to which we've dedicated the entire next chapter.

Chapter 15

The Evolution of the eXOgon Antenna

An Epiphany

By the fall of 1994, I was already a seasoned broadcast engineer, having accumulated 17 years of experience as Chief Engineer of KJNP, a 50,000 W directional AM station in interior Alaska, in addition to 22 years as a very active radio amateur.

I thought I knew something about radio by that time. As it turned out, I was just graduating kindergarten.

I received a somewhat mysterious phone call from a Dr. Alfred Y. Wong. Dr. Wong was the head of the UCLA plasma physics department, and was in the process of assembling a massive radio research facility about 25 miles east of Fairbanks, which became known as HIPAS Observatory. It was also to become the predecessor of the much more famous HAARP facility, which, as of this writing, is being reactivated after

Figure 15.1 — HIPAS Observatory from the air. Eight cross-polarized antennas graced the Alaskan wilderness like a great Steel Stonehenge. HIPAS was the predecessor to HAARP. It has been since decommissioned and dismantled.

a long, strange hiatus. But HAARP is another story altogether.

Dr. Wong had learned of my existence through the grapevine, solely because of my experience with high power vapor-phase-cooled transmitters. I had no physics credentials whatsoever, but I did know how to get high power transmitters up and running. As it turned out, HIPAS had eight of them, each with a power of 125,000 W, for a grand total of 1 megawatt of raw RF power…not counting about 17 dB of antenna gain. So I was hired as essentially an electrician and plumber.

The antennas for HIPAS were already in place when I came on board. The array was somewhat reminiscent of a giant steel Stonehenge, consisting of a central tower surrounded by seven other towers in an 800-foot-diameter circle. See **Figures 15.1** to **15.4**.

Atop each of these 60-foot towers was perched a pair of crossed dipoles; it didn't resemble any broadcast array I'd ever seen…or any ham radio station, for that matter.

I learned that these dipoles were *circularly polarized*, and that by selecting either right or left hand circular polarization, one could selectively energize different regions of the ionosphere. Naturally I had a lot of physics to learn, but the important point was that *circular polarization* was fundamental to how everything worked "up there." You could do a lot of very interesting things with CPOL that you couldn't do with "plain" (linear) polarization.

Figure 15.2 — One of the eight crossed dipoles of HIPAS. Each of these antennas had a set of traps to allow operation on 2.85 MHz and 4.53 MHz.

Figure 15.3 — Satellite view of the HIPAS antenna field. Aluminum irrigation pipe, 8 inches in diameter, was modified to serve as the transmission lines.

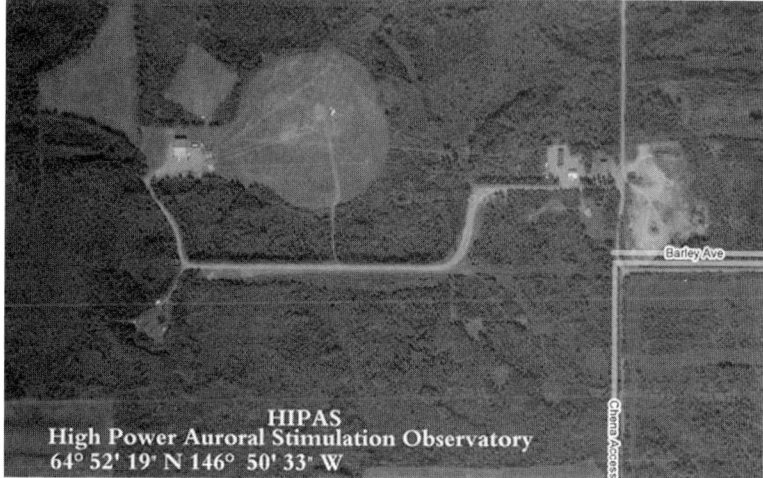

Figure 15.4 — A complementary receive array is barely visible in the lower right, just above the "t" in Observatory. It consists of eight crossed inverted V antennas with the same spacing and orientation as the transmitter array.

I would also soon learn that even if you started out with linear polarization, by the time the ionosphere was done with it, you had circular polarization, anyway. I began to wonder if any radio practitioners outside of the "Steel Stonehenge" had a clue about any of this. Why weren't any hams doing anything with any of this? Was I privy to some bizarre classified information?

A little poking around revealed that, no, we weren't the first. A *QST* article in 1940 described X and O propagation very clearly and concisely. Oddly enough, however, there had been no *subsequent* articles on the matter whatsoever. It was as if there was a massive conspiracy of silence about the matter in Amateur Radio circles. But this "enforced ignorance" was not to be found in the scientific community. X and O modes had always been taken for granted in the physics realm — even before radio. It had already been described in minute detail by the optics community.

After I had gotten all the transmitters running, which did take some doing, my next assignment at HIPAS was to build a complementary *receiving* array about a kilometer away from the main transmitter array. So out on the tundra behind the primary antenna farm (upon which some of the biggest blueberries I'd ever seen grew with abandon), I assembled a circular array of CPOL inverted Vs with basically the same dimensions as the main transmitting array, but for receiving purposes only. Each of these antennas was fed to a phase coherent receiver so we could do 3D direction finding of signals coming back from the ionosphere. We optimistically called this receiving array an HF Holograph. It didn't do *everything* it was supposed to do, but it was extremely educational.

What was astonishing to me (but evidently not to any of my physics colleagues) was how *well-defined* the polarization was of these skywave returning signals. If I had the *sense* reversed — for example, if I tried to receive an O mode signal with the receiving array set to X — the signal basically went away. It became screamingly obvious that this X and O mode business was not just a mere footnote, but that it was absolutely *central* to ionospheric propagation. In other words, it was *big*. For me personally, it was the most profound discovery I'd made in all my years of playing (and working) in radio.

Missing Ingredient

Most profound discoveries are usually hiding in plain sight. While circular polarization was neither mysterious nor arcane for most technically-minded radio amateurs (at least those reasonably versed in satellite communications), on the HF bands, circular polarization was unheard of. Admittedly, a CPOL antenna at HF frequencies takes up a bit more real estate than the "normal" kind of antenna, but beyond that, no hams that I knew of, outside of the ionospheric research community, had any *reason* to use CPOL…or even try it out.

Paul Graham, author of *Hackers and Painters*, a compendium of his excellent essays on entrepreneurialism and creativity, said that innovation often happens in clumps — if you really have a great idea, probably someone, or a few *someones*, are thinking the exact same thing. This is probably a bit hard on the ego, but it also means you're probably on the right track.

At HIPAS Observatory, Dr. Mike Trimpi, whom I've already mentioned in a previous chapter, sometime around 1996 was looking for a way of measuring the polarization of aurora generated noise. This required not only an antenna with good CPOL discrimination, but also one that was readily steerable. Up to that point, we had a vast collection of antennas at HIPAS that filled the former requirement, but not the latter. CPOL antennas at 2.85 MHz and 4.53 MHz (our primary "heater" frequencies) were on the large side, and more or less permanently fixed in an NVIS configuration. So a few of us technicians (I wasn't even the *main* guy on this project) came up with a pair of *active* hand-held circularly polarized turnstile antennas, one for each of the heater frequencies, that could be readily steered at the suspect auroral region. They worked admirably for the task at hand, but I had other ideas. These antennas showed in a very simple and convenient manner that HF signals were indeed circularly polarized. If we could clone these antennas and make them readily available for hams, we could at least get a significant number of hams at least *considering* the idea that X and O modes existed. Only one ingredient was missing — the ability to make antennas these cover a multitude of HF bands. It's incredibly easy to make a CPOL turnstile antenna on one frequency, but making one for multiple bands is a bit trickier.

The magic bullet that made this idea possible was a wideband 90° hybrid splitter that had just become available by Mini-Circuits, a company with whom I had become intimately knowledgeable during my tenure at HIPAS. Specifically, it is the JSPQ-65W chip (**Figure 15.5**), which has a precise 90° (quadrature) phase shift between two ports over a frequency range of 5 to 60 MHz. As far as I know, this is the *only* commercially available device that has these specifications. The device works amazingly well, and I hope it never becomes "unobtainium." While unable to obtain an actual schematic of this particular device, Jeremy Cortez, a most helpful application engineer with Mini-Circuits supplied me with a very detailed thesis on the design of such wideband networks, along with his assurance that this device will *not* become unobtainium in the foreseeable future.

Figure 15.5 — The Mini-Circuits JSPQ-65W chip which has a precise 90° (quadrature) phase shift between two ports over a frequency range of 5 to 60 MHz.

With the combination of this device and a pair of broadband, active, orthogonal dipoles, we managed to create a high-performance, compact, and convenient X-O antenna that any ham can duplicate.

What's in a Name?

After some brainstorming with a few of my entrepreneurial partners-in-crime, collectively known as AlasKit Educational and Scientific Resources, we came up with a name for the finished project, the eXOgon antenna. The capital X and O in the name refer to X and O modes, of course. We did obtain a registered trademark for the name eXOgon, but not a patent, despite some very strong exhortation to do so. We felt it far more important to get this device in the hands of a lot of hams than to try to make any money at it. We *may* apply for a *design* patent in the future, if we come up with a really attractive package that hams can't wait to get their hands on. In the meantime, you can make your very own eXOgon antenna and experiment to your heart's content with the reception of X and O modes. **Figures 15.6** through **15.10** show the eXOgon details, and we'll have more on this antenna in Chapter 22.

Figure 15.6 — Jeff Dumps holds the prototype eXOgon antenna, inspired by the HIPAS array. The antenna is constructed of electrical conduit with magnet wire running through the arms. The electronics is contained in the hub in the center of the arms.

Figure 15.7 — Schematic diagram of the eXOgon antenna. There are two identical preamps in the system, one for Northeast-Southwest and the other for Northwest-Southeast. The op-amps are AD8067 FET-input op-amps. The winding of T1 is quite forgiving, consisting of 6 trifilar turns on Amidon FT-50-43 cores. A number of different cores were tried with no measurable difference in performance.

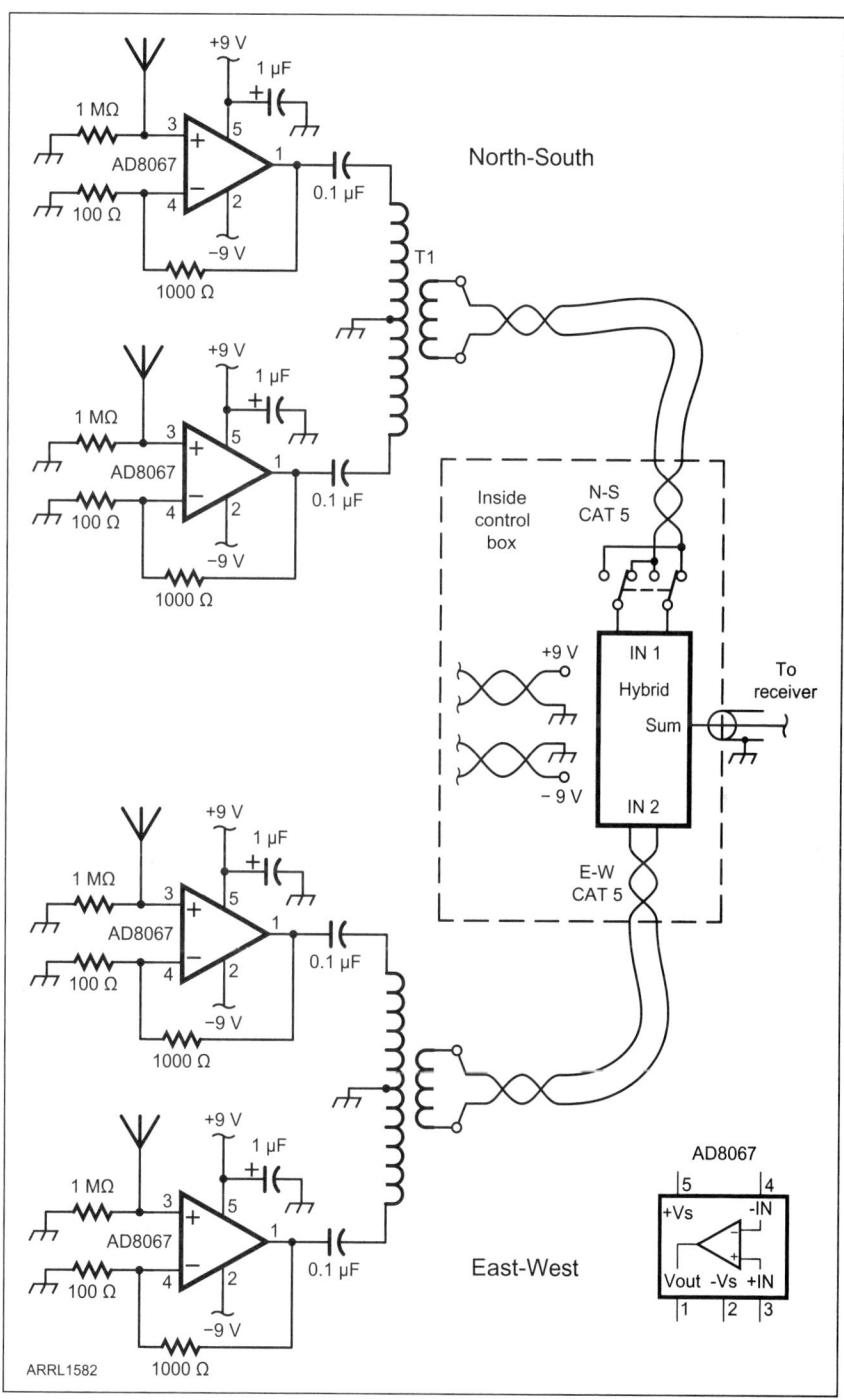

Figure 15.8 — Inside the eXOgon antenna hub.

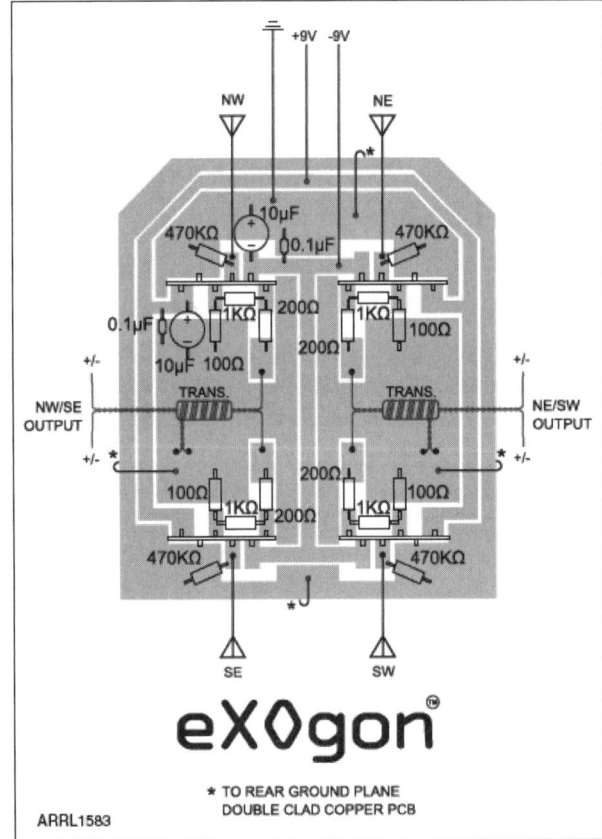

Figure 15.9 — PC board layout for the eXOgon antenna preamplifier.

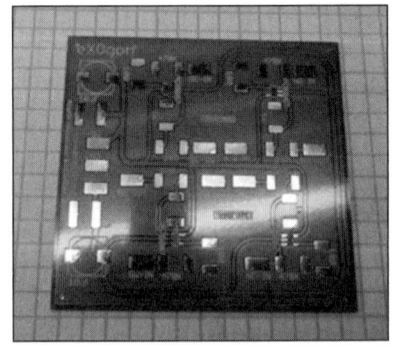

Figure 15.10 — Custom fabricated PC board to accommodate four AD8067 op amps, a handful of passive devices, and the push-pull output transformers.

Chapter 16
The Lowdown on LF

Well, isn't this interesting? I had *just* begun the joyfully arduous task of writing this chapter, when lo and behold, we received the long-anticipated news that the FCC had just approved *two new amateur bands*! This news is so new that the ink is still wet, even as I write this! (Well, it *would* be wet if we still used ink for such matters.)

The two new bands we acquired were on 630 meters and 2200 meters, the former being pedantically an MF (medium frequency) band, and the latter being an LF or "long wave" band.

Serendipitously, (at least for the sake of this book), nothing like these two new bands can so graphically and thoroughly demonstrate the benefits of active antenna technique and technology. Everything we've said up to this point on the matter is stretched to the logical limit when dealing with these two new bands.

The Long and Short of It

Our new 630 meter band is about four times the wavelength of our previous longest wave amateur band, 160 meters. That means that all things being equal, getting on 630 meters is about four times as hard as getting on 160. But all things are not equal. There are *severe* power limitations to our new allocations, generally 5 W ERP, and in some cases 1 W ERP. Transmit antenna efficiency is extremely low, and as such, most signals are going to be minuscule in strength. That's the bad news. The good news is that radio amateurs have a 12-year track record of successful experimentation with such restrictions. The ARRL 600 Meter Experimental Group website at **500kc.com** chronicles the exploits of the intrepid pioneers on this band. Their ceaseless efforts are solely responsible for us acquiring these new bands, and they are to be congratulated and honored appropriately. The best way we can honor them is to *get on the air and use these bands*!

Anything we say about the 630 meter band can be extrapolated to the 2200 meter band…if we multiply it by about three and a half. But let's focus on 630 meters first.

The Old and the New

Operating on frequencies well below the current AM broadcast band is not a new accomplishment for radio amateurs. Amateur Radio *began* down in the nether regions of the radio spectrum. What *is* new are the severe power restrictions. But we have some new weapons in our arsenals that our brethren of yore lacked — weak signal digital modes such as FT8, JT65, and JT9 and a host of others — all of which are capable of extracting the equivalent of the hiccup of a flea in the midst of a hurricane. And not to mention, new active antenna methods to aid and abet these amazing high-tech receiving widgets.

Amateur Radio can and will take place on these extremely challenging bands. Don't be daunted by the naysayers. And, especially, do not belittle our new allocations; it cannot be overstated how hard-won and victorious these new bands are. They open up exciting new avenues for experimentation, and will allow us once again to *define* just what the state of the radio art will be.

Little Ears

A great deal of wonderful information has been generated on the challenges and successes of transmitting on MW and LF bands. We refer you again to **500kc.com** for lots of great information on short but not *too too too* inefficient low frequency antennas. Our tireless experimenters have probably approached the limits to what can be done in terms of transmitted ERP within our legal restrictions. It goes pretty much without saying that further success on these new bands will be primarily determined by the receiving end of the matter.

As we hope we have conveyed in our chapter on the disappearing antenna, this should not be too much of a problem. Although, undoubtedly, a number of hams will rely on classic monster antennas like Beverages to receive our MF and LF signals, the majority of them will have to work with much smaller antennas. And the good news is, as we have shown, very small receiving antennas, relative to wavelength, can be immensely effective.

Ground Rules

With some possibly very rare exceptions, propagation on 630 meters will be entirely by ground wave. It is conceivable that on some occasions, modes such as chordal hop might occur on 630 meters, in which case signals could arrive from high angles. This is almost certainly never going to happen on 2200 meters. However, we do need to add the "almost" qualifier to the "certainly never," because, as Amateur Radio has historically and consistently shown, nothing is impossible. But it's certainly safe to say that we should concentrate our efforts on ground wave propagation, which greatly simplifies things. First of all, we can rule out much experimentation with horizontally polarized antennas, since ground wave propagated signals must be vertically polarized, as we have already shown. And as somewhat of a corollary to this, we can assume that such signals are going to be low or very-low angle signals, traveling along the surface of the earth.

Titanically Speaking

For more than a century, 500 kHz was the default ship-to-ship and ship-to-shore disaster frequency. The Titanic had spark transmitters on 500 kHz. This frequency range has remarkably reliable coverage over sea water. Semi-ancient mariners will tell you how 500 kHz was a worldwide party line for radio operators, giving reliable world-wide coverage day and night…even with the basically "deaf" receivers of the era. Now the situation is reversed. We will not have kilowatt spark transmitters on our new 630 meter band, but we will have receivers the Titanic crew would die for. Well, they died anyway, but it had nothing to do with radio technology…or lack thereof. That's another story for another book.

At any rate, 500 kHz has very low attenuation along sea water, and 2200 meters has even less. So, even though we must start out with minuscule signals, once they are launched, they tend to *stay* launched for considerable distances. This is where lower frequencies have a distinct advantage. Now, most of us aren't going to be operating our new bands on the high seas. But even though *dirt* is nowhere near as good a medium as the ocean, it's still pretty good for radio's nether regions.

DF Goodness

Because skywave propagation becomes less effective (primarily due to D-layer absorption) the further you go down in frequency, MF and LF signals tend to very accurately follow great circle paths. This is something that HF signals *don't* do, as our discussion of X and O rays shows. So, ground based direction finding methods are very reliable on our low bands.

As it turns out, very small tuned loop antennas, not only have high selectivity, but also have the best directivity. Antennas such as ferrite loopsticks have razor sharp frequency selectivity, and near perfect nulls… a great combination for direction finding along the Earth's surface.

The ferrite loopstick has a long track record with AM broadcasting reception, as well as on long wave. Now, most of us W/K-land hams probably haven't done much long wave broadcast listening, but most of our European partners-in-crime have — especially in Northern Europe, such as Finland and Sweden. One of the favorite weapons of the more urban European long wave folks seems to be *arrays* of ferrite loopsticks. (Laplander LF broadcast DXers, who have as much available real estate as the reindeer they chase around the Arctic, seem to be fonder of Beverages.) As we showed in our chapter on beamforming networks, this makes a whole lot of sense.

Hoarding

I don't know why ferrite should be a particularly expensive material, but larger chunks of the stuff seem to be so. There was a time when rather large ferrite rods were available for a song (and perhaps a dance thrown in), but they seem to be in short supply. After pricing an 8 inch long, 3/4-inch diameter ferrite rod for an unrelated science project, I choked a few chokes and gasped a few gasps, and decided I should start hoarding the stuff. Again, I have no idea *why* ferrite of any kind should be expensive, but I'm not a chemist.

I did find a wonderful unexpected source for ferrite rods…in the form of ferrite RF suppression beads. Normally these are designed to be slipped over coax cables to form current mode chokes. But you can super-glue a few of these end to end and come up with a pretty authoritative loopstick, and the gaping hole through the middle doesn't seem to have any detrimental effects.

Jack Smith, K8ZOA, derived a concise formula for the ideal length and diameter of a loopstick antenna for any particular purpose (see "Observations on Ferrite Rod Antennas," *QEX* Jul/Aug, 2008, pp 22-35). A useful way to think about this is the ferrite rod is the magnetic analog to a short electric wire antenna, so the longer the better! However, there are a few trends that are easily demonstrated without resorting to complex formulas. First of all, the permeability of the loopstick increases with the length of the rod. And it does seem to increase indefinitely as you lengthen the rod, confirmed by Smith's measurements, though progressively less dramatically as you add length to the rod. Second, a loopstick of any given length has the greatest inductance if the coil is wound at the center. In addition, the null is much more symmetrical if the coil is centered on the rod. I suppose this should be intuitive, but it's nice to actually measure such things (again, see Smith's article).

The diameter of the loopstick seems to have a much less dramatic effect. However, some good folks at **www.vlf.it**, many of whom are working with frequencies *much* lower than our two new ham bands, seem to suggest that skinny ferrites (high length to diameter ratio) are better. This seems to be inconsistent with the general formula that $ = performance, so if true, this is wonderful news. But I defer to the VLF specialists on this point.

If the Litz Fitz

There's a whole class of radio experimenters that even most Amateur Radio operators never encounter. You can meet many of them on the aforementioned site: **www.vlf.it**. I became acquainted with them a couple of decades ago when I was experimenting on the 1750 meter band (between around 160 – 175 kHz). There are some serious hard-core science dudes in this gang, and occasionally their knowledge creeps into the amateur ranks.

One of the unresolved discussions is the need, or lack thereof, of using Litz wire when winding loopstick antennas and others of their ilk. Litz wire is a flexible "wire rope" material that is made of many strands of copper wire, each strand being *insulated* from the other. The purpose of this is to reduce the skin effect, by increasing the total surface area for a given "rope" diameter. The math and formulas for calculating just how much improvement you get by using Litz wire are extremely complicated. And there are countless forms of Litz wire, using endless varieties of "weaves" and whatnot, each with unique radio frequency properties.

If you look at any AM portable radio, you will find a loopstick antenna wrapped with Litz wire. I've made a lot of loopstick antennas with normal single strand wire, and haven't seen any real difference that I could actually measure. However, this is probably a place for exploration, especially on our new 2200 meter band. It could be that the use of Litz wire has a major effect on performance. I honestly don't know. But, Litz wire is fairly inexpensive, so it might be worthwhile to play around with it so as to be able to contribute intelligently to the radio art.

More Ground Rules

Going back to some things we *do* know. The terminated loops we previously discussed, such as the pennant, flag, EWE, and the like, could also be scaled for MF and LF frequencies, and they would perform essentially the same as they would at HF. That's one of the real beautiful things about RF, as opposed to something like aerodynamics, where you have all kinds of non-scalable, non-linear parameters to deal with. However, although any antenna can be scaled to any frequency, not all antenna types are *practical* at all frequencies. This is why we don't see many parabolic dishes for the AM broadcast band.

Now the pennant and the flag antennas have, as one of their selling points, the case that they're "ground independent" at all frequencies in the sense that the ground is not part of the antenna circuit. In other words they may be deemed "complete antennas" without any ground connection. They do, however couple to ground (mutual impedance), and distant ground reflections impact the antenna patterns While ground independence may be true at the upper HF frequencies, it's debatable at the lower HF frequencies, and certainly not true at MF and LF.

It's Complex

When speaking of ground permittivity, most hams are oblivious to the fact that permittivity consists of a real part and an imaginary part. It is more accurately described as an *admittance* figure. The imaginary part of

---

Tickled

As discussed earlier, there are definite advantages to achieving all the selectivity possible right at the antenna. Of course, this goes exactly contrary to the concept of a wideband, active whip antenna designed to cover much or all of the HF spectrum. By their very nature, broadband whip antennas are susceptible to intermodulation problems, caused by mixing of multiple strong signals…or even worse, the mixing of a strong signal with a weak signal you are trying to copy. Only active devices and circuitry with exceptional dynamic range can be used with fully broadband antennas when they are in the presence of strong signals.

The use of highly selective passive components before the "front end" greatly reduces the requirements of the active devices, both by reducing wideband noise, and rejecting nearby frequencies. The downside of this, of course, is that such antennas need to be tuned. For a generation of hams used to having everything in the station "no-tune," it may seem burdensome to have to retune "everything" whenever you move a few kilohertz in frequency. Many new hams have never even encountered a *preselector* on a receiver or transceiver.

However, previous generations of hams actually *liked* the ritual of twisting knobs, yours truly being no exception. It gives you something to do while the band is opening up, at the very least.

One of the oldest methods of achieving extreme selectivity right "up front" was the regenerative receiver. A *tickler coil* was used to reinsert a small amount of positive feedback into the of the first RF amplifier (usually the *only* RF amplifier), along with the input signal arriving from the antenna. By careful adjustment of the tickler coil, the receiver could be set just on the *verge* of self-oscillation, where theoretically infinite gain and selectivity were possible. In reality, the gain and oscillation were somewhat less than infinite, but nevertheless, *extremely* high. It was not uncommon to achieve 100 dB of gain with a single RF amplifier tube, using regenerative methods.

the ground permittivity is proportional to conductivity (σ) in siemens per meter. (I prefer the old term *mho* to siemens, because it's easy to remember that with mo' mho, you have mo' current flow. So mho is mo' better. But since 1971, all our electrical units have to be named after dead science guys, and thus siemens won out over mho, which is Ohm in reverse.)

Regardless of the actual numbers involved, we know with a certainty that the degree of effect that the ground has on any radio signal is directly related to the wavelength. For an antenna to be unaffected by ground, it has to be much higher at low frequencies than it does at high frequencies. An antenna can usually be safely deemed as "in free space" if it's 10 wavelengths or so above ground. That's an awful high antenna on 2200 meters!

So the bottom line here is that we can't eliminate the effect of ground on any practical MF or LF antenna. The best we can hope to do is *take*

Traditional tickler coils used a small rotatable coil inside of the RF amplifier's main tank circuit. They were somewhat mechanically awkward, and tended to be quite touchy. Maximum coupling occurred when the tickler coil windings were parallel to the tank windings, and minimum when the tickler coil was perpendicular to the tank coil. (Larger versions of the tickler coil, known as *variometers* were used for transmitting purposes, before circuits like the ubiquitous pi-network became standard.)

Later receiver designs moved the regenerative RF amplifier "downstream" into the IF amplifier stage or stages, where it became known as the Q-multiplier. Q-multipliers were often added on to mediocre receivers to make them somewhat less mediocre.

A regenerative circuit can also be moved *upstream* from the classical regenerative scheme, to include the actual antenna itself in the positive feedback loop. When positive feedback is applied to a tuned loop, residual copper losses in the loop can be effectively reduced, further reducing noise and increasing selectivity. A small tickler coil can be inserted concentrically with a tuned loop, similar to the "driving" loops often used with small transmitting loops. A small sample of the RF after the first amplifier can be fed back to this tickler coil, with appropriate control. If the antenna is located outdoors, you will need to provide a second transmission line to drive the tickler. The optimum size and spacing of the tickler will have to be determined experimentally.

As in days of yore, you *must* be careful not to allow the system to go into self-oscillation, as this will also transmit quite nicely. This was one of the major problems with the widespread use of regenerative receivers, as squeals and whistles from nearby (and not so nearby) receivers could be heard all over the bands. However, this is a very effective way of achieving a lot of low-band gain and selectivity when other methods fail. It is recommended to use *just* the amount of feedback necessary to notice the improvement in selectivity, *not* on the ragged edge of oscillation. You already have plenty of gain downstream with any modern rig.

This is old technology that can really work well with state-of-the art methods.

advantage of the ground effects. And that's something we can actually do quite elegantly with the right kind of antenna.

Since MF and LF ground waves are so predictable, we can design antennas to take advantage of this predictability. As indicated at the beginning of this chapter, we want to optimize our antennas for vertically polarized low angle waves. What kind of antennas do this for us? Well, the Beverage is one, to be sure. However, to have any significant gain, it has to be huge. If we have no room for huge, we need to look elsewhere.

Narrowing the Field

One aspect of these new bands that we haven't mentioned is the fact that they're narrow…very narrow. This means that any communications mode will also have to be extremely narrow. Communications will necessarily be rather slow. The **500kc.com** folks relied heavily on QRSS signals — *very* slow CW, just a few characters a minute. Using *lock-in* amplifier techniques, you can copy a very slow CW signal well below the noise floor, and that is *without* any fancy digital signal processing.

Again, like anything else in radio, there is no free lunch. The price you pay for very long integration times, such as when using lock-in methods, is that you have to deal with very low data rates. However, for the purposes of experimenting with our low frequencies, where QRSS CW is the standard mode of operation, this is a reasonable price to pay.

Again, the only cost is *time*.

Since bandwidths in the new bands are dictated by the narrow band allocations themselves, we can augment this "deficiency" by even narrower antennas. How do you make an antenna super narrow? The same way you make *any* circuit super narrow — by the judicious application of positive feedback. Most hams of the properly seasoned variety remember the large variety of Q-multipliers that were available. A radio frequency amplifier *just* on the verge of oscillation has both infinite gain and infinite selectivity. There's nothing to prevent you from applying a little positive feedback to an already high-Q tuned loop (such as a loopstick antenna) and achieving even *higher Q*. What we have now that the Q-multipliers of yore didn't is easily controllable gain devices, such as logarithmic op-amps. It's fairly easy to build a tuned circuit right on the verge of oscillation, but never crossing over, with modern RF devices.

Untickled

I first learned of this concept from Dr. Mike Trimpi, (famous among plasma physicists for the Trimpi Effect) while I was working at HIPAS Observatory. It is well known that the use of feedback resistors in any amplifier circuit, op-amp or otherwise, creates noise. Several methods of "lossless feedback" using either directional couplers or transformers have been used to overcome such noise. The most notable is the lossless Norton (or Norton-Podell) amplifier, US patent 3,891,934 shown in **Figure 16.A**. Dr. Trimpi suggested that *radiation resistance*, being lossless and noiseless in nature, could be used as the feedback element in an op-amp circuit, thus tailoring the frequency response of the antenna and reducing noise in one fell swoop. The antenna resulting from this concept quite resembled the K9AY loop, but with an extra coupling loop thrown in to couple the radiation resistance back to the input of the active circuitry. This is another technique that merits a lot more experimentation. In Chapter 22 we will present a rudimentary "Trimpi Loop" construction project for the curious ham.

This variation on the tickler theme uses *negative* feedback on the antenna to broaden the bandwidth of a loop antenna, and "calm things down" if necessary. The benefits of a super-sharp loop are lost when using negative feedback, but for a lot of scientific investigation, it's more important to have predictable and controlled gain than to have a lot of selectivity. You just want to be sure you don't *add* any unnecessary noise to the system, and this lossless feedback method assures that, by not adding any resistance in the feedback path. Again, this is a method that has been seldom used in Amateur Radio circles, and lends itself well to experimentation.

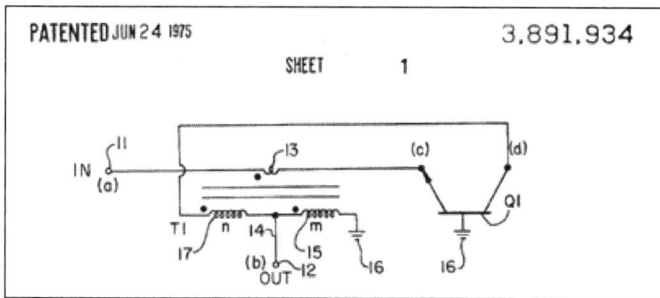

Figure 16.A — The lossless Norton-Podell amplifier.

Chapter 17

The Random Wire

We were a bit reluctant to include an entire chapter on the random wire antenna, because it is somewhat antithetical to the trajectory of this book, which is presenting exciting and novel antennas using the latest technology.

However, after some thought, it became apparent that there's a lot more going on in the typical random wire than most folks suspect. And, it is also the most effective and easy way of getting *something* on the air for countless new short wave listeners (SWLs) and hams. The fact of the matter is that the random wire just plain works a lot more than it doesn't, and so deserves more than a parenthetical mention.

What It Is

We should discriminate between a random wire and a long wire, although there are often some overlapping terms and uses. A long wire antenna *can* be random, but can also be extremely well engineered. Typically a wire antenna is deemed a long wire if it is at least several wavelengths long and exhibits some definable directivity and gain over a dipole. V-beam antennas and rhombics for example, fall well into the engineered long-wire category.

Conversely, a random length of wire can, in many cases, exceed the length of many engineered long wire antennas. So, for the sake of this discussion, we will define a random wire antenna as any random length of wire with an undetermined or unknown radiation pattern. So a short whip antenna, such as our now-famous disappearing antenna, falls into the random wire category.

Another qualifier is that the random wire is generally considered to be end fed, but this is not absolutely necessary to qualify.

Because You Can

A great deal of online Amateur Radio discussion (if this can be trusted for actual data) seems to center on the trials and tribulations of using an end-fed random wire for transmitting. I could spend half of this book deconstructing the *whys* and the *why nots* of using an end fed random wire antenna for transmitting. Issues such as "RF in the shack" are quite likely to arise when attempting to transmit with a random end fed wire. But despite the drawbacks, many hams, including myself, have successfully used these devices on occasion.

However, when such an antenna is used for receiving only, most of those issues disappear. If you are a new ham or SWL, the best reason for using a haphazard random wire antenna is *because you can.*

Compare and Contrast

Although we've spent the bulk of this book investigating the practical differences between receiving antennas and transmitting antennas, it's a good idea to touch base with their similarities from time to time. And nowhere do the similarities between receiving and transmitting antennas become as clear as when you model them. In fact, when setting up an antenna for modeling with a program such as *NEC2* (in any of its variations), there is nothing anywhere to differentiate between a transmitting and receiving antenna. Nothing whatsoever. There's no clue in a *NEC* modeled antenna which role it's actually playing. And this is a useful thing to know.

But before diving headlong into antenna modeling, let's remember that no antenna modeling program is a substitute for thinking. So let's look at some of the more obvious characteristics of a piece of wire.

The Straight Dope

Straight pieces of wire are the easiest to analyze. This is not to suggest that there is anything special or advantageous to using a straight piece of wire over a bent one. Many highly effective receiving antennas are not only random length, but also random shape. As I write this, I am in the process of coaching a long-lapsed radio amateur back into the hobby. (We were both licensed about 45 years ago, but somehow life got in his way, and he put the hobby on hold after his Novice license expired. As for me, my own QSL card states, I've been "…continuously hamming since 1972.")

Neither here nor there, said "retread" has a more-than-decent-sized lot, and in an enviably quiet electrical location. However, his house is not only very tall, but of such an architectural configuration as to make it nearly impossible to access the roof from any direction. Either the original architect was extremely anti-ham, or just being silly. The house is, additionally, on one corner of a rather long and skinny lot, which slopes severely toward a lake. And furthermore, despite no hard corroborating evidence, he is absolutely certain that any of his neighbors would be horrified by anything that looks like an antenna.

The bottom line is that any wire he can put up will take an extremely convoluted path around walls, porches, fences, and perhaps an accommodating tree or two. It appears that this will also be his transmitting antenna, which may *indeed* present some challenging engineering issues. However, let's save that for another day and look at why just about any wire he tosses up will probably be a fair receiving antenna.

The End Game

A straight piece of wire in free space cannot receive a radio wave arriving end on, nor transmit a signal in that direction. This is because the electric field has no component in the direction of the wire. Well, we could be a bit pedantic here and say that if the wire has any *thickness* we can induce electrons to wobble back and forth across the gauge of the wire. For all practical purposes, a wire antenna can be considered infinitely thin, at least for this discussion. There is a "cone of silence" directly off the end of any wire antenna, either for transmitting or receiving.

In addition, the impedance at the very end of a wire in free space is *utterly undefined.* How can this be? Well, to have an impedance you have to have a circuit…or at least a complete component of some sort with *two terminals.* Impedance is defined as *between* two points in a circuit. If you only have one point, there's nothing to measure the impedance *to.*

As is the case in many purely mathematical contexts, *undefined* is often synonymous with *infinite.* We say that the slope of a vertical line is *undefined*, but we also know it is infinite. Likewise, the impedance at the end of a wire in free space can also be considered infinite. When using antenna modeling programs, you'll usually throw some kind of error if you place a voltage generator at the very end of a wire. And, when it comes to real, physical antennas, any long thin wire in any environment resembling free space will have an impedance too high to match with any practical device. Fortunately, we don't usually erect antennas in such an ideal environment, which is actually a good thing. Hold that thought; we'll return to it soon.

Let us return to our cone of silence for a moment. Yes, the cone of silence is real, and it can have real consequences under some conditions. This fact, however, is often taken to ludicrous extremes. Large numbers of hams dread the possibility that their wire antennas may have a null in some important direction. Well, this might be the case if your shack were in outer space. However, at any reasonable height above ground, most HF antennas are fairly non-directional. Ground reflection does a very good job of filling in the cone of silence. But don't take my word for it; let's do some modeling with my favorite modeling program *4nec2*. Let's pick some fairly random parameters, too. Let's say that life and real estate dictates that your only choice for a receiving antenna is a 60 foot wire running between the peak of your roof (at say 25 feet) and a tree in the back yard. Just to make things not *too* random, we'll make the wire horizontal. We also have to find a way of connecting the wire to our receiver, so we'll add a 22-foot vertical drop line from the peak of the roof to your ground floor window. Let's see what this thing does on 40 meters.

We'll start out with a basic geometric model of the antenna (**Figure 17.1**). Notice that we have no actual ground connection, but we do have an extensive *average* ground. Let's run a far-field pattern and inspect a few of the actual electrical details (**Figure 17.2**).

First we see that the SWR (referenced to a 50 Ω transmission line) is 4372:1. This indicates that this is probably not something you want to connect directly to your standard HF transceiver, at least for transmitting purposes. The input impedance is 310 – *j*8226 Ω. While this is very high, it is not quite infinite. The reason for this is that the ground underneath the feed point of the antenna does act somewhat as the second terminal, as far as impedance is concerned. It's not a *great* terminal, but at least

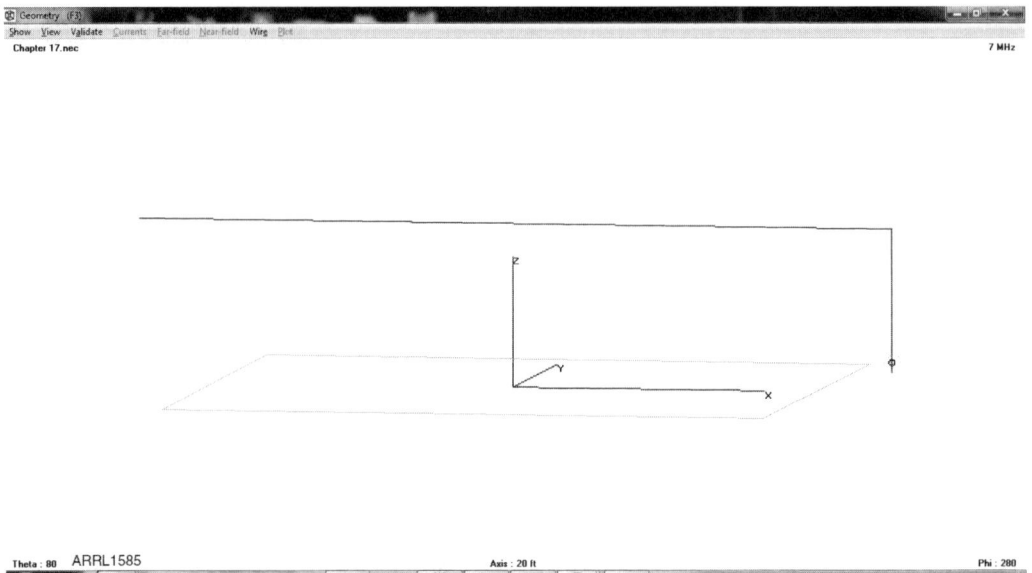

Figure 17.1 — **Basic geometric model of a random wire antenna.**

Figure 17.2 —
Electrical details of our
modeled antenna.

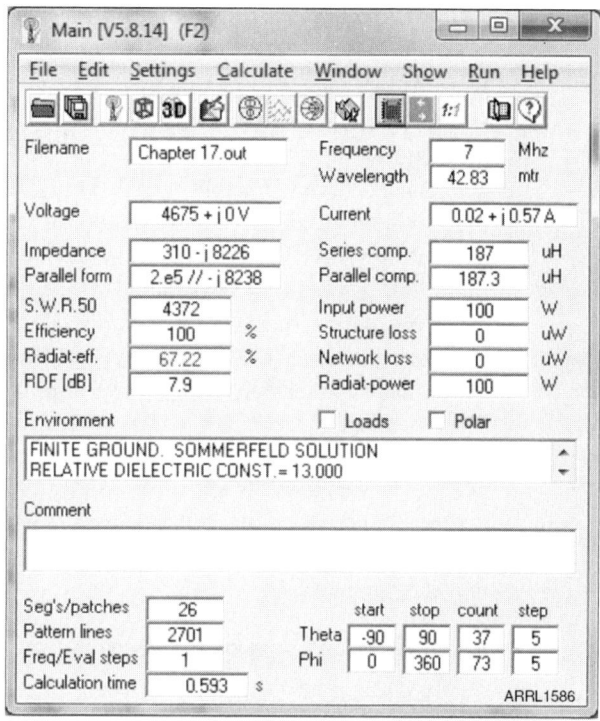

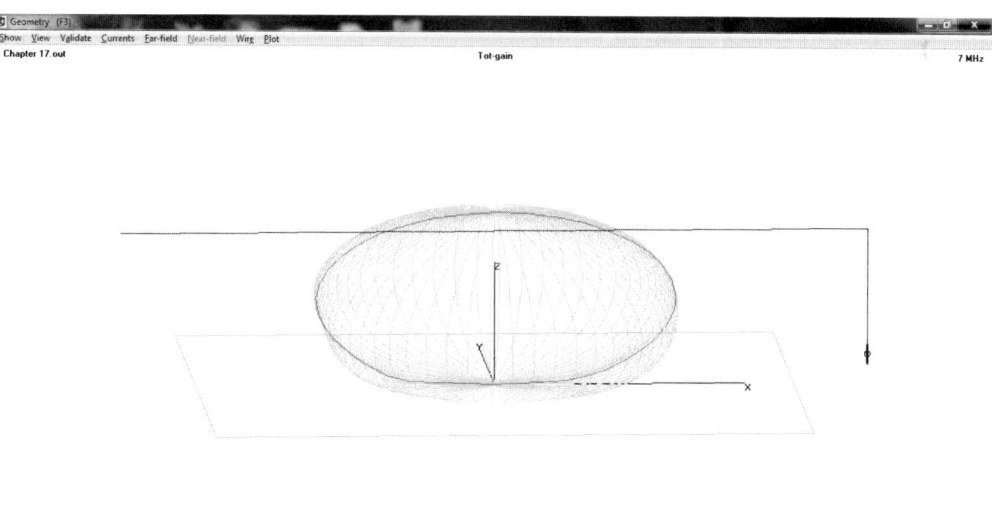

Figure 17.3 — Radiation pattern of our modeled antenna.

The Random Wire   17.5

it exists to the extent that our *NEC* model can actually produce a value. This would not be the case if we modeled it without some kind of ground. Interestingly enough, even though the antenna is nominally a resonant length, it has an extremely high value of capacitive reactance, to the tune of 8226 Ω.

What accounts for this extremely high capacitive reactance? Well, the fact is we have inserted our generator in the very last minuscule segment of our wire. The return terminal is actually just the lower half of that minuscule segment. We basically have a generator with one leg flapping in the breeze. Neither *NEC* nor reality takes too kindly to this situation, but it's surprisingly easy to resolve. But before we do that, let's look at the radiation pattern of this atrocity (**Figure 17.3**). Interestingly enough, the radiation pattern is nowhere near as atrocious as the impedance. It is essentially non-directional, and utterly devoid of the dreaded "cone of silence" at either end of the wire. In fact, this turns out to be a perfectly acceptable receiving antenna, as far as radiation pattern is concerned. We can take an even closer look at this pattern, by using the 3D function of *4nec2* (**Figure 17.4**).

Figure 17.4 — 3D radiation pattern of our modeled antenna.

Figure 17.5 — Adding three extra feet of wire to the model and attaching it to our ground.

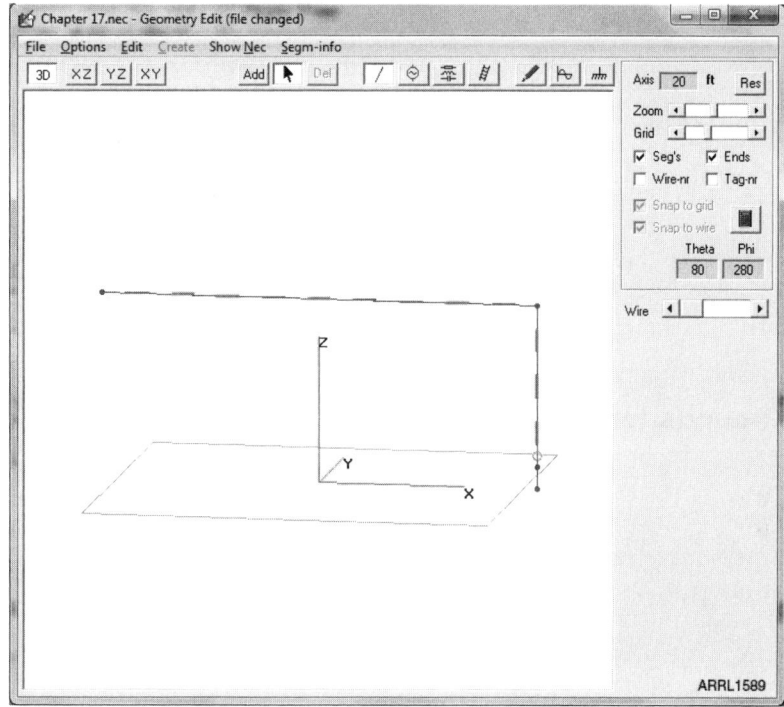

We have previously shown, hopefully to your satisfaction, that impedance matching is not particularly crucial for most modern day receivers. But if we're on the obsessive side, we could add 8226 Ω of inductive reactance in series with the feed point to make our antenna look reasonably resonant, in which case it will look like 310 Ω. This isn't a great match, but probably better than you need. A lot of boat anchor receivers have single wire input terminals that are designed for impedances on the order of 300 to 600 Ω. And, unless you're a real whippersnapper, you've probably seen a knob labeled "antenna trim" on a general coverage receiver or two. The antenna trim was typically just a series capacitor to help adjust random antennas such as the one in our example. Generations of hams have used random wire antennas with no more of a nod to impedance matching than the beloved "antenna trim" twiddle-widget.

Let's make just one modification to our random wire, to make it, well, a lot less random. Let's give the "bottom" of our signal generator somewhere to go, by adding three extra feet of wire and attaching it to our ground, as imperfect as that is (**Figure 17.5**). Looking at the impedance

(**Figure 17.6**), we see we've already made a drastic improvement. Now the SWR is down to a "mere" 56:1. However, we haven't laid out anything resembling a ground system; we've merely poked the end of the wire into the lawn, so to speak.

But as we see in **Figure 17.7**, there is no discernible change in the radiation pattern. If nothing else, this shows quite graphically that there is no connection whatsoever to radiation pattern and impedance…for nearly any kind of antenna. This is a truth that if universally understood would eliminate a lot of confusion about antennas. And, more generally, this shows that, at least at HF frequencies, there's little to dread with regard to unexpected deep nulls, especially for simple antennas.

Grounds for Reception

I suspect that most experienced hams are presently appalled and aghast by my assertion that poking the end of a wire into the ground constitutes anything resembling an RF ground — as well they should be. I merely wanted to make the case that a true end-fed wire is really an aberration; it doesn't really exist.

But we now should address the issue of just how good a ground needs to be for receiving purposes. Much has been written about the need for effective ground radio systems for any kind of low-band *transmitting* success, whether using verticals or not. I defer to the giants of the literature in this regard, such as *ON4UN's Low Band DXing*, for the final word on effective grounding systems. I see no need to reiterate the case here, except for one interesting twist.

Charcoal and Other Alchemy

I thoroughly enjoy reading some of the earliest literature on radio experimentation, especially with regard to crystal radios for the AM broadcast band (and sometimes longer wavelengths). Much was made of the fact that half of any antenna was under the ground. All kinds of alchemy was described in the efforts to obtain a good low-resistance ground, some of which called for insertion of large quantities of vile potions into the Earth that would certainly raise the curiosity, if not the ire of the Environmental Protection Agency or other similar entities, if attempted in this century. This was in an era when antenna and *Marconi* antenna were nearly synonymous.

The benefits of good low-resistance grounds for crystal radios were indisputable, although the *reasons* were not quite as clearly understood. It's actually quite simple, in retrospect. Nearly all crystal radios at the

Figure 17.6 — Electrical details of the antenna model with three feet of wire added. Now the SWR is down to a 56:1.

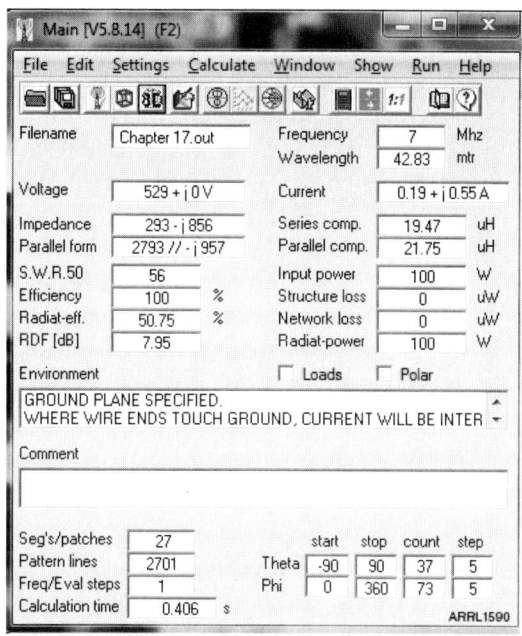

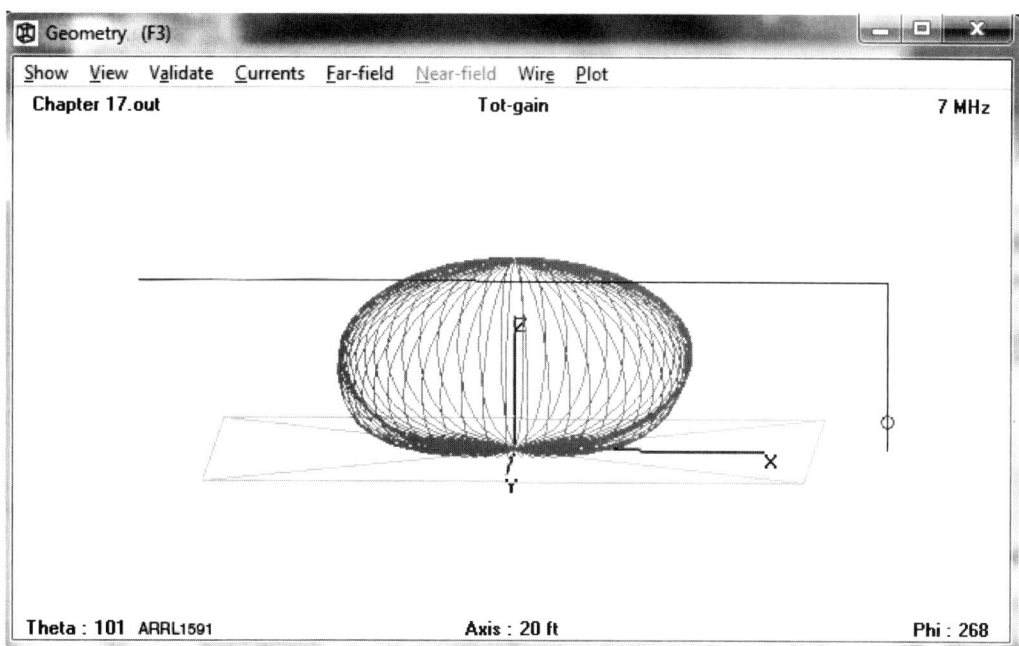

Figure 17.7 — Radiation pattern of the antenna modeled in Figure 17.5. There is no discernible change in the radiation pattern compared to the original version in Figure 17.3

time incorporated *series resonant* antenna circuits, of which the aboveground wire (or conglomeration of wires) and the underground "terminal" were the necessary components, along with the required antenna coil incorporated in the receiver. In a series resonant circuit, the Q is limited by the series resistance, and ground resistance was the primary source of this resistance. It was absolutely crucial to bring this to the lowest value possible in order to preserve the entire system Q.

On the other hand, a *parallel* resonant antenna, such as a frame loop, has a Q dependent only on the internal resistance of the wire. It did (does) not rely on a low resistance ground — or any ground at all — to maintain its Q, or to even function. And Q is the real secret to crystal radio sensitivity, as well as selectivity. Later developments, such as the "loose coupler" acknowledged the secondary importance of a low loss ground for receiving effectiveness.

So the question is, for general purpose reception in the 21st century, is there any benefit to paving one's back yard with copper? Or injecting your subterranean real estate with cubic yards of copper chloride and charcoal? Probably not. As recent modeling algorithms, as well as a great deal of practical experience indicate, signal strength, both for transmitting *and* receiving, is largely determined by ground conditions *far* beyond your antenna's immediate surroundings. For transmission purposes, yes, you want to have a good robust return path for your near field currents, which dictates some kind of radial system for vertical antennas. (And they certainly don't hurt for dipoles either!) But ground conditions many wavelengths from a low-band vertical antenna have a profound effect on the far field signal strength, and for *receiving* purposes, the "far out" ground-field conditions are probably the only ones that matter. There probably isn't enough charcoal and copper chloride available to make any difference. However, we shouldn't be too dogmatic about this; it is certainly open to experimentation, especially on our two new low frequency bands.

Virtual Antenna Length

As we have suggested in several places, engineering is the art of compromise. You have to work with what you've got. When it comes to antennas, we realize that a lot of hams "ain't got much" when it comes to options for putting up sky wires. Where there's a will, there's a way. However, sometimes that way is a rather bent and convoluted one.

Although the straight wire is the easiest to analyze, it isn't always the simplest to install. Great numbers of hams have to rely on bent and convoluted wires…both for transmitting and receiving. One of the most common questions asked on a number of ham radio internet forums is, "What will happen if I bend my wire around the corner of my house," or, "Is an inverted V as good as a dipole?" or "My antenna consists of a wire wandering three hundred feet through a forest. Is that a problem?"

Many years ago, I came up with sort of a rule of left big toe to answer this question. It's not precise, but it seems to work in a large number of situations. I call it the Virtual Antenna Length, which pretty much works for any randomly wandering wire. Find the two most distant points on your woods-wandering wire. Draw a straight line between those points. This straight line is your effective antenna, and the length of this line is your Virtual Antenna Length (VAL). The radiation pattern of this effective antenna will be very close to what your actual antenna's pattern is, as well as the radiation resistance. The simplest case of this is the inverted V. Both the input impedance and radiation pattern are nearly identical to what you'd have if you strung a dipole between the two existing end points of the inverted V.

Now, I suspect that someone or a chorus of several "someones" will require that I rigorously prove the validity of VAL for any conceivable case. I can't. But it seems to be fairly intuitive as to why it works as well as it does, if intuition has any credibility.

The Active Random Wire

We have one weapon in our arsenal that crystal radio operators of ages past did not possess. And that is the near-infinite-impedance active device. As we described earlier, the *antenna trimmer* was a partial answer to random impedance, end fed wires. But we can take this a step farther by using modern active devices to interface with the bitter end of an end fed random wire. As in the case of the magical disappearing super-short antenna, we really don't have to attempt to match impedance. We can use our random wire as a simple voltage probe, and use our active device to convert that voltage to a current, allowing us to use the directional advantage (and possible gain) of a semi-long wire over a short whip, but with the desirable wideband characteristics of the latter.

We will revisit the random wire later on as we explore *diversity reception* methods. We will describe a very clever way of using a very random wire to obtain surprisingly good space diversity.

The End of the Matter

The end-fed half-wave antenna (EFHW) has undergone a tremendous resurgence of interest for *transmitting* purposes, especially for limited space installations. In fact there's a whole Facebook page dedicated to amateur EFHW antennas: **www.facebook.com/groups/EndFedHalfWaveAntennas**. A wide variety of EFHW hardware is available from outfits such as Palomar Engineers, as well: **palomar-engineers.com/tech-support/tech-topics/best-end-fed-antenna-for-ham-and-swl**.

It's safe to say that any of this equipment designed for transmitting purposes is certainly adequate for receiving purposes, though possibly a bit of overkill for the purpose. While a number of these prefabricated antennas are specifically designed as *half wave* antennas, they are certainly suitable for use as *long wire* antennas as well. There's no hard and fast definition of what constitutes a long wire, other than it's generally deemed to be a wavelength or longer at the frequency of operation. Naturally an EFHW antenna for 40 meters is a long wire on 10 meters.

While any EFHW antenna needs a counterpoise of some sort for transmitting, (contrary to some advertising literature, but see **www.w8ji.com/end-fed_1_2_wave_matching_system_end%20feed.htm**), a receiving antenna is not subject to such restraints…necessarily. This is not to say that a receiving antenna will *not* benefit from an extensive counterpoise system. Perusing through the earliest literature of radio, one is struck with the time and effort dedicated to exotic grounding and counterpoising methods used for merely receiving AM broadcasts! There must have been *some* measurable results for the blood, sweat, and tears applied. At least it certainly can't hurt. If you've got the space and the copper, use it!

BALUNsing Act

One of the things that has allowed the EFHW antenna popular again is the ready availability of efficient toroidal core transformers, both of the powdered iron and the ferrite variety. Such gems didn't exist during the heyday of the Zepp antenna, which is the most celebrated incarnation of the EFHW antenna. The plethora of EFHW transformers on the market today (often erroneously referred to as *baluns* in this application) make life a lot simpler for the space-strapped ham…both for receiving *and* for transmitting.

A typical EFHW transformer has a nominal impedance ratio of 9:1. This is based on a semi-educated guess that a typical half-wave antenna at "typical" height above "typical" ground is going to be somewhere in the region of 450 Ω. From personal experience, helping a number of hams get up and running with EFHWs, I've found that this is a fairly good guess, resulting in less than 2:1 SWR on the average at the operating frequency.

For strictly *receiving* purposes, however, I believe a much higher transformer ratio is desirable. I tend to default to a 64:1 ratio, presenting an impedance of 3200 Ω to the antenna's end. This allows the antenna to operate with almost no discernible loading on the end, preserving its near free-space radiation pattern, which can be of benefit when using the nulls for noise reduction. Remember we aren't shooting for maximum power transfer in the receiving realm, just a good signal-to-noise ratio.

Chapter 18
Arrays and Beamforming Networks

The phased array antenna has been around for a long time. In fact, phased array antennas of a wide variety were in common use long before the invention of the Yagi-Uda antenna, which is normally the sort of antenna most hams think of first when talking about directional antennas.

AM broadcasting was where phased arrays were first used on any large scale. And to this day, the phased array is the *only* type of directional AM antenna approved by the FCC for commercial broadcasting purposes. The reason for this is simple. With a phased array, you have full control over the antenna currents and radiation pattern at all times. With a Yagi-Uda or other parasitic array, once the antenna is designed and constructed, it's difficult to adjust to compensate for changes in the environment. Parasitic arrays can have tremendous gain for a given size, but they also tend to be a bit touchy.

Another advantage of the phased array antenna is that it can be steered electrically, whereas a conventional Yagi-Uda has to be steered mechanically.

Both

None of the above in any way suggests that parasitic arrays and phased arrays are mutually exclusive. Large phased arrays of Yagi-Udas are frequently used in applications such as moonbounce, as well as countless scientific applications. One of my first assignments at HIPAS Observatory was to build a UHF radar comprised of a square array of 16 17-element Yagis. It was an immensely successful project, and allowed us to get some truly original data on field-aligned irregularities in the auroral region. Twenty years after the fact, there's been a tremendous resurgence of interest in this phenomenon; I was privileged to have been somewhat on the ground floor.

Probably the most famous and spectacular specimen of a phased array is the 360-element HAARP (High Frequency Active Aurora Research Project) array in Gakona, Alaska (**Figure 18.1**). I spent many a day crawling around under this

Figure 18.1 — The 360-element HAARP (High Frequency Active Aurora Research Project) array in Gakona, Alaska.

array in the not-too-distant past, and will be doing so again in the near future.

Phased arrays can be made of *any* kind of antenna, although the design is much simpler if the individual elements are of the same type, and of fairly uniform spacing. However, scientific instruments such as the VLBA (Very Large Baseline Array) radiotelescope use phased array methods to combine the signals from widely flung and irregularly spaced elements into a single coherent pattern. These elements are separated by thousands of miles.

Running Interference

All directional antennas — whether a phased array or a parasitic array or even a parabolic dish — rely on wave interference to create the final pattern. By carefully and deliberately combining the RF signals from a collection of antenna elements, we can create just about any kind of pattern desired. One thing that you can do with a phased array that you absolutely cannot do with a parasitic array is come up with a perfect null. Many hams, and even professional radio folks, are astonished to learn that even a parabolic dish has a large back lobe! So any antenna that exhibits a perfect null in some direction will have some phased array method involved.

Phased array design is one area where reciprocity applies completely and absolutely. And this also implies that you can model such designs reliably and accurately, whether the end result is intended for transmitting or receiving.

A Subtle Difference

Although "phased array" and "beamforming networks" are terms that are often used interchangeably, there is a subtle difference. Technically, beamforming is what you *do* with a phased array, and focuses primarily on the actual phasing and combining hardware. But, perhaps, more interestingly, the modern concept of beamforming is the ability to derive several different radiation patterns from the same physical antenna array *simultaneously*. Naturally, the question that comes to mind is *why would any ham in his right mind want to do such a thing?* The obvious answer is, most hams aren't in their right minds to begin with. But beyond this, advanced beamforming networks are invaluable in applications such as radio direction finding, and a whole raft of interesting scientific applications. Combined with modern SDR receiving methods, clever beamforming methods can give you the equivalent of a room full of radios with just one radio.

The Butler Did It

One of the neat projects I helped put together for HIPAS Observatory (High Power Auroral Stimulation — the predecessor to HAARP) was an eight channel *riometer*. The name riometer is derived from Relative Ionospheric Opacity, which is an indication of ionospheric absorption. It's measured by actually looking at cosmic noise in the 30 to 50 MHz range. The device was actually invented in Alaska during the 1957 International Geophysical Year (IGY), which was a watershed event for ionospheric research — some 40 years before I got involved.

Our particular riometer consisted of eight collinear 10-meter dipoles, with the array oriented East and West such that we could see the diurnal (daily) changes on cosmic noise as the Earth rotated. Our array yielded eight different directional beams simultaneously, so we could see the cosmic noise source (purportedly the center of the universe) sweep across the array as the Earth rotated.

Probably the best way to demonstrate the concept of beamforming is by exploring the Butler Matrix (**Figure 18.2**). You can see there are

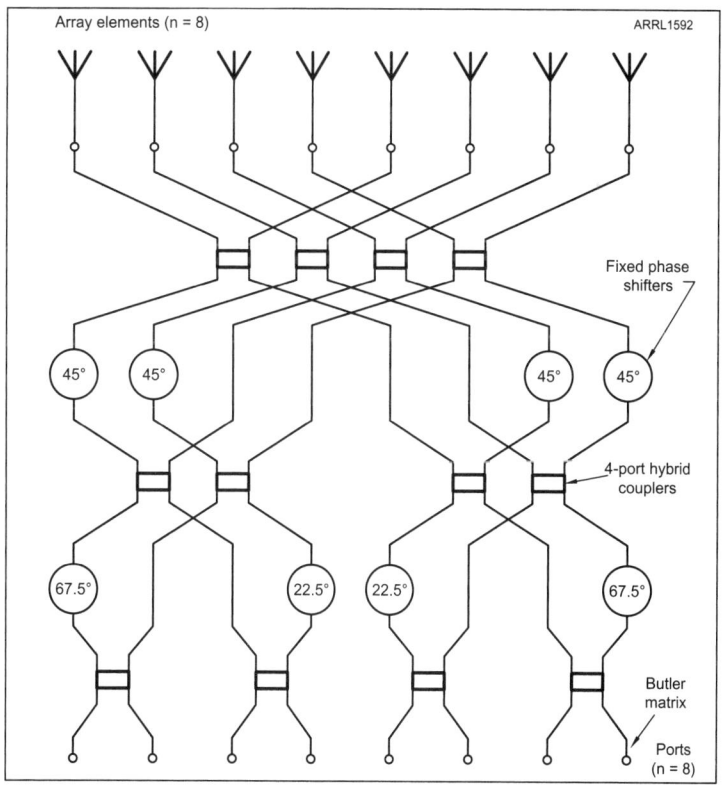

Figure 18.2 — Eight-channel Butler Array for HIPAS riometer.

12 90-degree hybrid combiners, and a number of fixed phase shift networks, which were really nothing but properly-cut sections of coaxial cable. Each of the output ports produced a beam from a different angle relative to the collinear dipole array. Originally, we had eight separate receivers, one connected to each output port of the matrix. Later on, we cobbled together a multiplexer circuit so all the outputs were inserted into a different time slot, but with just one receiver. SDR receive dongles have gotten so cheap lately that it's probably more cost effective to put eight of them together than to build a multiplexer like we did. Technological advancement can be wonderful.

At any rate, you probably won't have a need for such a complex Butler Matrix; we presented this just to demonstrate the concept. Your antenna array and associated Butler Matrix can be made as large or as small as you like (although only in powers-of-two increments). In other words, a "normal" Butler Matrix can only work with 2, 4, 8, 16, 32, 64…element arrays. There are other similar matrices that can work with odd numbers of antennas as well, but we can't "officially" call them Butler Matrices.

Piecemeal

Since we mentioned multiplexing, specifically *time division multiplexing,* it might be good to expand on the topic a bit, now that mortal hams actually have access to the technology to make it practical.

Normally, when forming a beam using phased array methods (whether simple or fancy), you combine all the signals in question at the same time. For example, if you have two stacked Yagis and you want to combine the signals for additional gain (in one direction), you simply combine the signals with a simple coaxial T connector. The signals algebraically add in the T (if you do everything right). You have one signal coming out of the T which is the composite of the individual elements' outputs.

But what if you only have one Yagi? Can you make a phased array with just one antenna…or is this just crazy talk? The answer is, yes you can! What if we receive a signal from, for instance, a Yagi, and store that signal. Now we move that Yagi to some other location and aim it at the same source, and then store that signal. Is it conceivable that we could take these two stored signals, compensate somehow for the time difference between our two samples, and then recombine them to get a new pattern? Indeed we can, and this is the principle behind the synthetic aperture antenna.

The earliest well-known implementation of a synthetic aperture antenna was the Wullenweber or "Elephant Cage" antenna used for worldwide radio direction finding during the Cold War (**Figure 18.3**). At one time, the Wullenweber was highly classified, but now it's all public domain — not to suggest that you'll want to build your own in your back yard. But you never know. The amazing thing about the Wullenweber

Figure 18.3 — The earliest well-known implementation of a synthetic aperture antenna was the Wullenweber or "Elephant Cage" antenna used for worldwide radio direction finding during the Cold War.

was that it was able to do all kinds of time division multiplexing using the completely analog methods available at the time. At any rate, the Elephant Cage created a virtual moving antenna that rapidly orbited around a thousand foot circle, and then reassembled the received signal samples to give an astounding amount of information about a distant radio source. A more recent amateur implementation of the Wullenweber general idea was the 2-meter DoppleScAnt, described in May 1978 *QST*.

Some Tidbits About Large Phased Arrays

Although the emphasis of this book is on HF and lower frequency antennas, where large arrays (in terms of number of elements) will be rather uncommon, there are some interesting universal principles to keep in mind when devising phased arrays for any frequency.

Rule #1: The closer the elements are in a phased array, the cleaner the pattern will be.

Rule #2: The farther apart the elements are spaced in an array, the higher the gain (up to a point).

Rule #3: Equal power to all elements in a phased array results in the highest gain, but also the strongest side lobes.

Rule #4: The lowest sidelobes for a large phased array occurs when the power distribution is tapered or "shaded." Greatest power should be fed to the inner elements of the array, with progressively less power toward the edges. This principle applies to antennas as diverse as dipole arrays to parabolic dishes!

Synthetic Aperture Techniques

Here is a topic for the truly geeky radio experimenter, but I suspect this, like many other cutting edge ham technologies of late, might soon become mainstream. In concept, it is quite simple, but the actual implementation of it is rather advanced. We include this here for the curious.

When creating a beamforming antenna, we create a pattern by combining the phase and amplitude of a number of antenna elements (such as dipoles). Take the case of a very long array of collinear dipoles, receiving a signal from a point source of radiation at a great distance. We know that we can locate the source of radiation by combining the phase and amplitude outputs of each of the dipoles in a proper fashion. If the point source is perfectly broadside to the antenna, we know the output signals of each element will be in phase. If the signal source is off to one side, there will be a progressive phase shift along the array.

But building very long arrays takes up a lot of space. What if, instead of building a huge collinear array of dipoles, we were to simply take *one* dipole, and drag it along the original antenna line at a consistent rate? We could take a "snapshot" of phase and amplitude at each subsequent position of our dipole, and store that data. Then, after the last snapshot is taken, we can combine the signals together, compensating the phase for the change in position, and thus reproduce the same pattern as the fixed collinear antenna. Pretty slick, eh? Of course, this requires having some memory, and it requires that the signal source is stationary during the transit of our movable dipole.

As bizarre as this method may seem, it is standard practice in instruments such as synthetic aperture radar (SAR) satellites, which are used to obtain extremely high resolution images of the Earth's geographic features. The principle is the same; we artificially *synthesize* a very large stationary antenna by means of a much smaller *moving* antenna.

The increasingly popular Reverse Beacon Network (RBN) would be a very handy backbone for implementing synthetic aperture techniques, as well, especially for investigating deep space radio sources. The RBN backbone would allow hams spread all across the country to combine signals from countless moderate sized antennas and come up with what amounts to one really *big* antenna.

Something to think about.

Rule #5: Efficiency and gain of a large array are two different things (Gain = Directivity × Efficiency).

Rule #2 is most interesting and revealing. I worked for a few years on Air Force TPS-77 medium range surveillance radars. These radars use a large phased array panel antenna with dozens of very closely spaced dipole elements. Because the dipoles are so close to each other, there is a great deal of mutual coupling between the elements. Each element had a directional coupler and a dummy load, which dissipates a full 50% of the power presented to the element…this was the power coupled by the adjacent elements. But the pattern was extremely clean, a requirement that fully justified the 50% loss of efficiency over a wide spaced array!

The HAARP array is similar, but not quite as extreme. Likewise, there is a great deal of mutual coupling between the turnstile elements. However, rather than just burning up the coupled power in a bunch of dummy loads, an elaborate energy recovering system delays and reinserts the coupled power into alternate array elements. But the point is, regardless of the technology involved, you can't bypass the physics. Clean arrays *will* be closely coupled.

Rule #5 is especially applicable to receiving antennas, and we've discussed this concept in different terms from the very beginning of this tome. Efficiency is important when you're transmitting; you don't want to waste too much precious transmitter power heating up earthworms or dummy loads. For receiving purposes, efficiency of little significance. The amount of energy you are extracting from the "ether" in the receiving process is minuscule compared to what's being transmitted. A few percentage points of efficiency difference is like an eyedropper in the ocean. Whatever gives you the best signal-to-noise ratio for the prevailing conditions is what you want. And often what you want is a really, really deep null in some direction, in order to reduce noise or interference. And this is where the phased array and its associated beam-forming potential stand out. Transmitting and receiving are not reciprocal processes (this has nothing to do with the Theory of Reciprocity). In transmitting we care about maximum efficiency and gain. In receiving we care about the maximum signal-to-noise ratio.

Front to Back

Noise and interference are two very different things, but they have one common trait: you don't want them. In all our previous discussion of signal-to-noise ratio, we have considered more or less random noise sources, whether man made, as from a leaky pole transformer, or natural, as in the case of cosmic or thermal agitation noise.

Adjusting Phasing with Your Antenna Tuner

Over the past several decades, antenna tuners have been thoroughly analyzed and evaluated. There are numerous wonderful designs and equally numerous not-so-wonderful designs. Practically all the discussion has centered on the matching range and efficiency of the various types of tuners in their transmitting roles. However, virtually none of the discussion has focused on the *phase response* of the typical antenna tuner — and for good reason. In most applications, the phase response of an antenna tuner is irrelevant. When it comes to phased arrays, however, the phase response of an antenna tuner can be a valuable asset.

In particular, the common T-type antenna tuner is most interesting in this regard. A typical amateur type T-network consists of two series variable capacitors, typically labeled TRANSMITTER and ANTENNA, and a shunt inductor (generally switched) between the two capacitors and ground.

When matching a 50 Ω transmitter to a 50 Ω resistive load, the two series capacitors will have identical values. And under *typical* conditions, there will be a phase shift through the tuner of *about* 180 electrical degrees.

As is the case with any three-component tuner, an impedance match can be achieved with multiple combinations of settings. (This is not the case with a simple L-network, in which there is only one possible impedance matching solution). While the "multiple choice" solution for a typical antenna tuner can be problematic at times, for the purpose of antenna phasing adjustment it can be highly beneficial. As it turns out, the T network is capable of adjusting the impedance match and phase shift *independently*.

If we have an antenna that's closely matched to a 50 Ω transmission line, we don't normally need an antenna tuner. However, we can insert the T-network onto the line, adjust the two series capacitors to equal values, and then adjust the L for a match. There will be numerous different ratios of L/C that will achieve a 50 Ω match, but each of these will have a different phase shift.

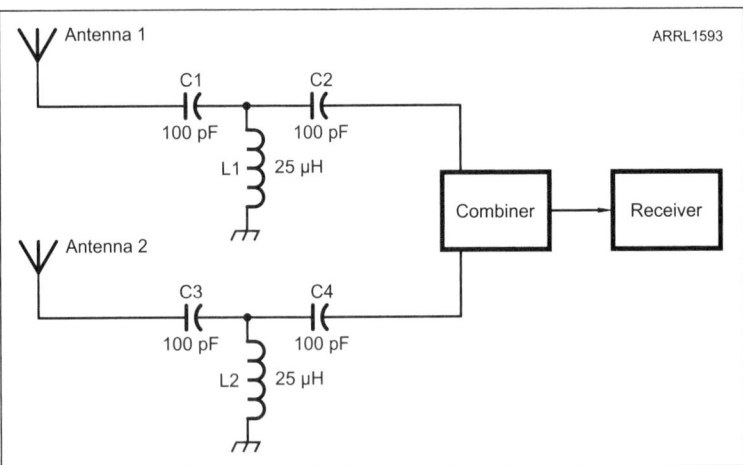

Figure 18.A — Incorporating T-networks in a two-element phased array to use the networks for fine adjustment of the relative phasing.

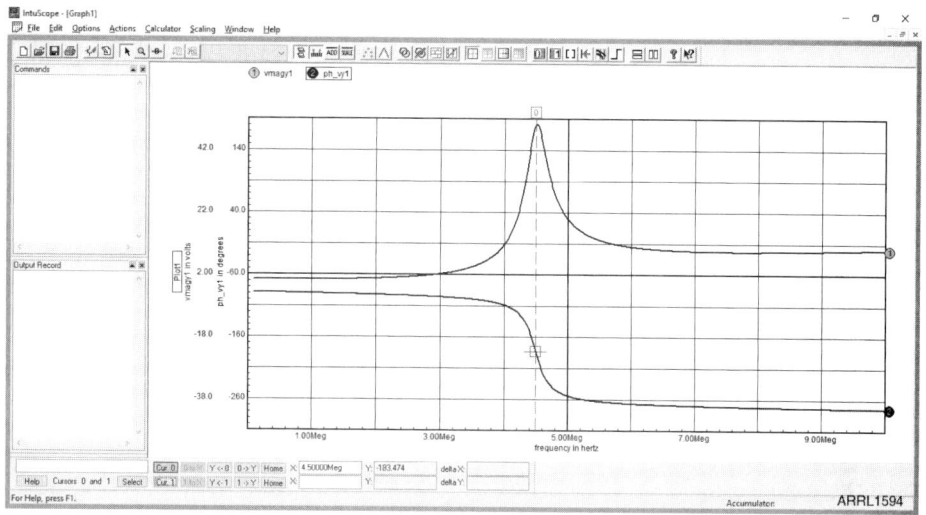

Figure 18.B — Phase shift of T-network with equal source and load impedance.

If we incorporate one or more T-networks in a phased array, we can use the networks for fine adjustment of the relative phasing, all while maintaining a 50 Ω impedance match. **Figure 18.A** shows how this idea would be implemented for a two-element phased array. Ideally, C1 and C2 would be ganged together, as would C3 and C4. By the way, MFJ makes a differential T tuner with a roller inductor, in which the two series capacitors are ganged in a differential mode. A slight mechanical tweak will put these two capacitors in a common mode. This makes a very versatile phase shifting network…though a bit pricey if you need to use several of them. However, low cost receiving versions can be constructed easily and inexpensively.

Figure 18.B shows the phase shift of a T-network with equal source and load impedance. The components shown for the T-network in **Figure 18.C** are only representative; in this case the phase shift is about 180°. Different L/C ratios will have different phase shifts while still maintaining a 50 Ω match.

Before implementing this in a receiving array, it is recommended that you develop a chart of phase shift for various L/C ratios for your particular tuner or tuners. This is a bit tedious to begin with, but will give you a very versatile phasing system when it's all done. I used a dual-trace digital storage oscilloscope to develop the phase figures, but you can use simple Lissajous patterns on a garden variety oscilloscope to perform the task as well.

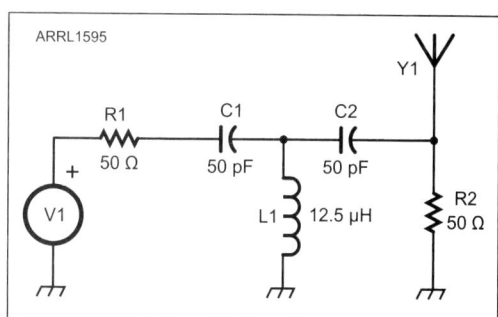

Figure 18.C — Component values used to demonstrate phase shift.

Arrays and Beamforming Networks **18.9**

Interference, on the other hand, is any unwanted signal, and in ham radio terms, it means any signal we don't want to communicate with *at that time*. That 40 dB over S-9 DX station that you've been trying to work for the past nine years is not interference while you're working him. But once you've made the contact, and you've moved up a few kilohertz, his splatter *is* interference. Well, it may not be his splatter that's causing you the grief; it could be just as likely generated in your receiver that has insufficient selectivity to handle the hurricane of RF that's pouring in through your antenna terminals.

In any case, one of the best ways to deal with strong interference is with a receiving antenna exhibiting a large front-to-back (F/B), or large front-to-side, or large front-to-something ratio. If you've done any Yagi modeling or actual construction, you know that optimum forward gain never coincides with maximum front-to-back ratio. Now, how important is front-to-back ratio when transmitting? Probably not much — I'd say not at all. Your main goal is to get as much forward gain as possible, even if that results in a lot of energy going somewhere else, too — like off the back of the beam. Can you see why this might be a justification for using separate transmitting and receiving antennas, even if they're both Yagis? You can have your receiving antenna optimized for best F/B and your transmitting antenna designed for the best "death ray" performance, regardless of sidelobes or other artifacts.

Stacking the Odds

Most HF operators don't pay much attention to the vertical angle of arrival of radio signals. If they did, we'd see a lot more azimuth-elevation (az/el) rotator schemes on HF antennas. How often do you see that?

Even if they did use az/el rotators, it probably wouldn't be terribly effective, except perhaps on the highest HF bands. Most moderate-sized HF beams have fairly wide "fans" in the vertical plane. On the other hand, a pair of stacked Yagis with adjustable phasing can exhibit much greater vertical steering range...not to mention being a whole lot faster. A pair of stacked Yagis has a more vertically "squashed" pattern than a single Yagi (as does a cubical quad, by the way). The relatively few hams that *do* use stacked beams typically use them to obtain a stronger low-angle lobe primarily by feeding them exactly in phase. But, by using essentially the same hardware *and* a little phase adjustment, this same narrow beam can be made vertically agile as well. This can result in not only greater received signal strength, but potential interference reduction, as well.

Another use is to fill in elevation angle nulls caused by ground reflections. The null (and peak) angles are related (very simply!) to the antenna height, so selecting between or combining signals from different heights helps.

Chapter 19

Powering Your Active Antenna

We have now arrived at the most boring chapter in the book. There's absolutely nothing exciting in this chapter, nor anything really original, but it's absolutely necessary to discuss the topic at hand.

An active antenna is one that has an active device…an amplifier. And all amplifiers need a source of dc power to operate. Most people think that dc is nowhere near as intriguing as RF, but without our boring dc, we wouldn't have much for our active antennas to work with.

The cleanest source of dc power available is a battery. With battery power, you don't have to worry about ac hum, many different kinds of feedback, and noise. The only problem with batteries is that you have to charge them or replace them.

At HIPAS Observatory, we had all kinds of battery powered instruments in the field, ranging from ELF to optics, some of these lasting for months or even years with little or no attention. And since then, battery technology has gotten orders of magnitude better. But no matter how reliable a battery might be, we can almost guarantee that it will be in a most inaccessible location and will fail at the most inconvenient time. So, generally speaking, we will want to look at some viable alternative to powering our active antennas — especially if they're going to be outdoors.

Warts and All

It seems that "wall warts," you know, those plug-in wall transformer power supply blocks used for everything from cell phones to electric pianos, grow on trees. They are immensely plentiful, cheap, and utterly non-standard. Every imaginable voltage, polarity, and output connector is available.

Some of them are impeccably clean from an electrical standpoint, and some of them are abominable sources of noise from dc to daylight. The only way to know for sure is to test them out. That being said, they are immensely suitable for powering a number of active antennas. But you have to do your homework first.

Outside

The very nature of an active antenna requires that the active element be located right at the antenna. If the antenna is some distance from the ham shack, this needs to be addressed. Running an extra dc power line from a wall transformer or other dc supply in the shack out to the antenna normally works fine — except that you need an extra run of copper. As you have undoubtedly discovered by now, nothing uses more wires than our "wireless" hobby. It may be inconvenient to run an extra cable out there and give one more thing for the yard moose to trip over or get tangled up in.

Biased

One of the tried-and-true methods for getting dc power out to some low-power remote RF device is by means of the coaxial cable itself. This is achieved through the agency of a "power inserter" sometimes known as a *bias T*. You also need a "power extractor" at the far end — another bias T. The bias T is nothing but a choke and a capacitor, wired so as to direct dc down one leg and RF down the other (**Figure 19.1**). At the shack end, the bias T can be incorporated as part of the power supply; at the antenna end, it can be incorporated as part of the amplifier. Or they can be entirely separate blocks. However, the prefabricated bias Ts used for cable TV and home satellite systems won't work for HF, as the inductance is nowhere near large enough, nor is the capacitance. So, for our purposes, we should build our own HF bias Ts, which is not a difficult job at all.

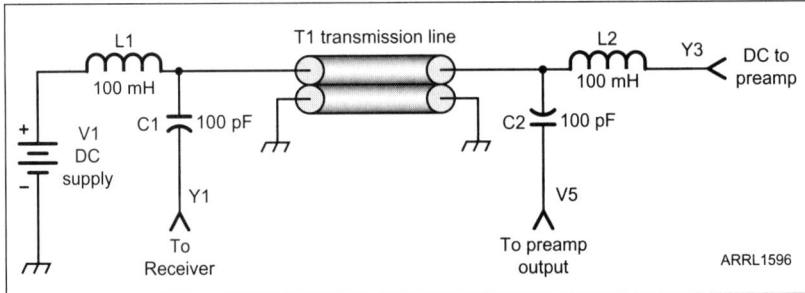

Figure 19.1 — Typical dc insertion/extraction (bias T) concept. Component values are typical for most HF/MF purposes. The two parallel transmission lines shown as T1 don't imply that you need two tranmission lines. This is simply indicating that the method can be used with any tranmission line, balanced or unbalanced. If using CAT5 or similar multi-pair cable, it is essential that any unused pairs are grounded, or high RF losses in the used pairs will result.

Baluns and Their Kin

In most cases it's an advantage to totally isolate the function of an antenna from that of a transmission line. The presence of a transmission line in the radiated field of an antenna is seldom beneficial, and can often be quite detrimental to the performance of the antenna. The topic of baluns and related issues is thoroughly covered in a lot of ham radio literature, including the *ARRL Handbook* and the *ARRL Antenna Book*. Most of the criteria for proper antenna/transmission line isolation are similar for transmitting and receiving purposes. However, there can be even more stringent requirements for transmission line isolation when it comes to receiving antennas. For example, precise direction finding requires that the pattern of the antenna is as close to ideal as possible.

The broadband balun (balanced-to-unbalanced) toroidal transformer is an effective and familiar item for most radio amateurs. However, most configurations of such transformers provide no path for dc between the transmission line and the antenna, or worse, present a dc short to the input side (and sometimes the output side). If you anticipate powering your receiving antenna/preamplifier through the transmission line, such transformers are unsuitable.

A Choking Matter

One fairly universal means of isolating the transmission line from an active antenna is by means of a "brute force" choke on the transmission line. Depending on the frequency of operation, the choke can consist of a few turns of transmission line near the feed point, or the use of one or more ferrite beads inserted around the transmission line, just "downstream" of the receiving antenna. This is the method I use on the eXOgon antenna. The brute force choke will have no effect on the dc path, and in most cases is extremely effective. It *is* important, however, to tailor the choke properly to the frequency range of operation. Some dimensions of coils can end up creating a *series* resonant condition with the remainder of the transmission line, which will render the entire scheme ineffective. On the other hand, a *parallel* resonant choke balun can work much more effectively than a non-resonant one, over a narrow range of frequencies.

Geometry and Symmetry

Surprisingly enough, it's perfectly possible for a transmission line to completely mess up an antenna radiation pattern even if a *perfect* isolation transformer or choke is installed. The best balun in the world can't do you any good if you *re-introduce* RF onto the transmission line *downstream* of the balun. (This is equally applicable in either a transmitting or receiving scenario.) An example of this is a dipole with a transmission line brought off parallel to one leg of the dipole. A balun won't do *anything* to reduce transmission line currents by such coupling. So, for an isolation scheme to be effective, the transmission line must be brought away perpendicular from the feed point. In most cases, this should be for at least 1/4 wave. As a general rule, if a transmission/antenna line system are *mechanically symmetrical*, you will probably have the most effective isolation.

Obsessive

Another alternative to this whole "balun-cing" act is to keep *everything* balanced. This requires the use of balanced transmission lines, such as twin-lead, open wire (ladder line), or Ethernet cable. (I use the latter quite effectively on the eXOgon.) The use of a completely balanced system will assure no interference between transmission line and antenna, as long as the rule of mechanical symmetry, discussed above, is complied with. For the truly obsessive direction finder, there is no viable alternative to a totally balanced system from antenna to receiver. Since you won't have any transformers to deal with, powering your antenna through the transmission line is not an issue.

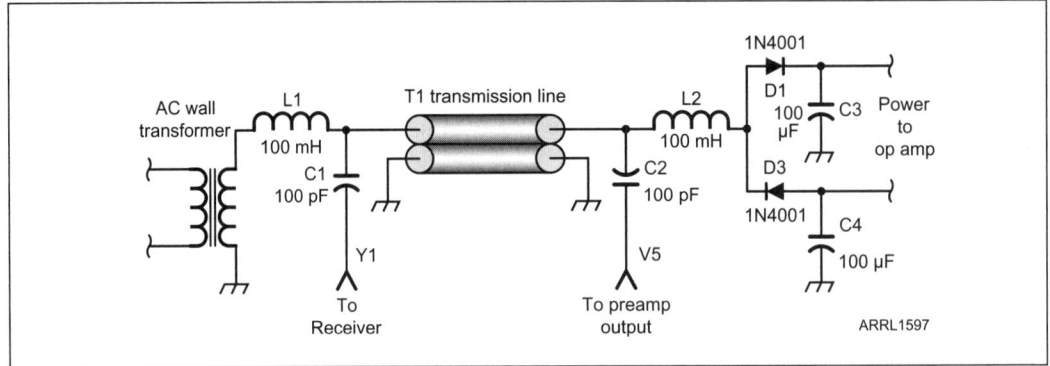

Figure 19.2 — Powering a bipolar op amp through transmission line, using an ac wall transformer is inexpensive and effective. Values shown are typical for MF/HF operation. See the Figure 19.1 caption for more information about the transmission line T1.

Bipolar

However…most of the active devices described so far in this book require *bipolar* power supplies (separate positive and negative supplies). While any op amp *can* be coerced to working with a single-ended power supply, many of them don't particularly care for it, especially the AD8067 which features so prominently in this work. We strongly recommend using bipolar power supplies for all active devices that prefer it.

So, this presents a dilemma. How do we send positive and negative dc voltage down the center conductor of a piece of coax?

We don't. We send ac down the coax (by means of a slightly tweaked bias T) and then *rectify* it at the antenna end with two back to back diodes, followed by two filter capacitors (**Figure 19.2**). This is a method I have used for many years with a variety of active antennas with great results.

You do want to be sure you have enough filter capacitance, since you're only using half wave rectification for each "pole" of the supply. But since these op-amps use so little current, the ripple will not be measurable with any reasonable amount of filter capacitance. If you really are obsessive, you can build choke-input filters after the rectifiers, but I've never encountered a case where this is necessary.

I've never had a problem with diode "hash" interfering with the desired signal, but if that *does* present a problem, you can use appropriate bypass capacitors across the diodes to squash it. I always use very robust decoupling on my active devices anyway, which is always a good idea for any low noise RF circuitry.

Figure 19.3 shows a couple of bias-Ts that I use.

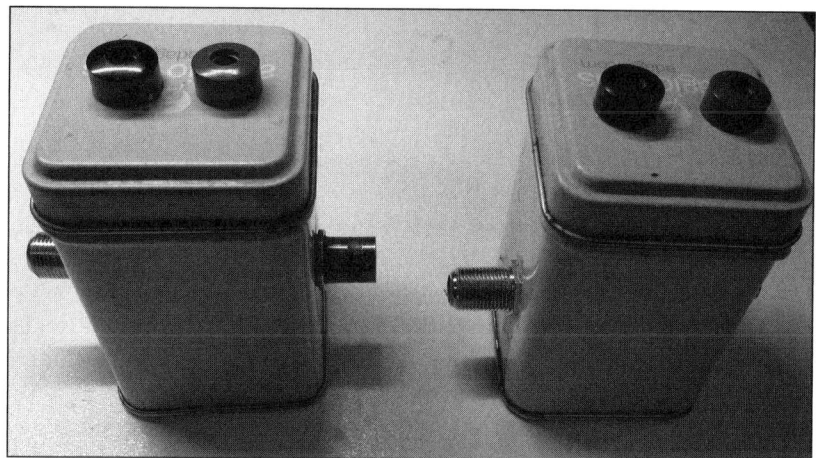

Figure 19.3 — A pair of bias "teas." Tea tins make handy cabinets for lots of projects, and they're well shielded. Jenny Tse of Sipping Streams Tea House always makes sure I have a steady supply of tea tins (and tea!). Banana jacks on top are used for power insertion and extraction; appropriate RF connectors are used for the transmission line connection and receiver/preamplifier connections.

Active Device Protection

It's probably safe to say that the majority of the readers here are licensed radio amateurs, or at least intending to be so. However, we don't want to leave out the large numbers of shortwave listeners (SWLs) for whom everything we've discussed so far is equally appropriate. In my long ham radio experience, I have discovered that technically, licensed radio amateurs have nothing over hard core SWLs when it comes to making effective receiving antennas.

It goes without saying that in most cases the active antenna is going to be in some kind of proximity to a transmitting antenna, some with considerable RF power radiating. We need to accommodate this environment when deploying any active antenna, since, for the most part, they contain sensitive semiconductor devices.

Field effect transistors (FETs) are especially vulnerable to strong RF fields, as well as electrostatic discharges (ESD). Although most modern FET devices do incorporate some internal ESD protection that can handle most normal situations, an antenna, being an antenna, is particularly vulnerable to lightning and strong RF fields. It should be noted that you don't need a direct lightning strike to cause severe damage to electronic and electrical equipment. The collapsing magnetic field from a relatively distant lightning bolt can induce huge currents in anything resembling a wire, which translates into extremely high voltages. I can tell some war stories of damage done to AM broadcast equipment from distant lightning strikes; and some of them are quite impressive.

As I stated in *The Opus of Amateur Radio Knowledge and Lore*, despite all our best engineering efforts to mitigate lightning damage, if a lightning bolt has your name on it, about all you can do is duck. That being said, here are a few things that can somewhat eliminate the vulnerability of active antenna electronics in the field.

There should be a dc path to ground, even if an extremely high resistance one, between any FET gate and incoming antenna connection. The very nature of an active antenna requires an extremely high gate impedance, so you don't have the option of placing a low resistance between gate and ground, which *would* offer a lot more protection. However, a value of 1 MΩ or more can prevent the slow buildup of static charges that would result if the gate were entirely floating. I use 1 MΩ "bleeder" resistors in my eXOgon antennas and so far I haven't seen any deterioration of performance, though, in principle, I'd really like to eliminate those resistors.

If you are in a "lightning alley" it might be a good idea to devise some means of disconnecting any antenna from an active device, and grounding said active device when lightning is in the area. This could, however, be more trouble than it's worth, as even the most elegant devices described in this book are only a couple of dollars.

As far as RF overload/toasting is concerned, by far the most effective measure is to install any active antenna as far away from any transmitting antenna as possible. The inverse square law of radio attenuation is really your friend in this regard. The amount of RF entering your sensitive preamplifier drops dramatically with distance from a transmitting source. Cross polarization, if possible, is also an extremely effective protection. If you transmit a horizontal signal, and use an active vertical whip, the amount of energy being couple is going to be inconsequential even at legal power limits, with any reasonable degree of physical separation. In most installations there is no need to devise any extraordinary measures to protect an active antenna from RF exposure at normal amateur power levels.

Transceivers

I did a rather extensive survey of present day HF transceivers to find out how many were accommodating of separate receiving antenna. I am delighted to report that nearly all mid-range and high end transceivers by the "Big Three" manufacturers have provisions for separate receive antennas, either by means of physical jumpers or front panel controls. Only the "bottom end" transceivers of any of the popular manufacturers lacked any provision for separate receiving antennas. Of course, if you're using "separates," this whole issue is irrelevant. Your transmitter and transmitting antenna and your receiver and receiving antenna are two entirely different entities.

For the more obsessive, there is the option of building an elaborate transmit-receive (TR) sequencing system, to provide absolute protection of your active antennas or other external preamplifiers. The latest few editions of the *ARRL Handbook* describe such a sophisticated sequencing box. It's a little overkill for my taste, but that option is certainly available.

Coaxing Your Signal into the Shack

As explained in the introduction to the disappearing antenna, a voltage probe "whip" *must* be directly connected to the preamplifier. The capacitive reactance of a very short whip is so high that any parallel capacitance, such as that presented by even the highest quality coaxial cables, will essentially short out the amplifier input.

However, what happens *after* the preamplifier is a different story. Any number of options are available for getting the signal from the low impedance output of the preamplifier to the shack. There is no need to obsess about transmission line impedance, or impedance matching. Cable TV coax, such as RG-6 or RG-59, which is 75 Ω, is in abundant supply and more than suitable for long runs between your active antenna and your receiver.

As I described in Chapter 15, the eXOgon antenna uses CAT5 cable for this task. I've run hundreds of feet of this with no significant ill effects. Of course, open wire feed line is viable, as well, though for receiving purposes, coaxial cables may experience a little less noise pickup. Again, if things are working properly, the noise figure is already established by the preamplifier before it even gets to the transmission line. Active antennas are picky about what goes *into* the amplifier, but are very forgiving of whatever follows, at least at HF frequencies.

Chapter 20

Diversity Methods

Probably no technique available to the radio amateur can accommodate the vagaries of HF radio propagation better than *diversity reception*. Diversity is actually a *family* of reception methods, and includes *frequency diversity*, *polarization diversity*, and *spatial diversity*.

While diversity reception methods have been known for many decades, they have had very limited application in Amateur Radio circles. For the sake of this discussion, we will bypass *frequency diversity*, as that is primarily the domain of VHF, UHF, and microwave communications. We will focus on spatial diversity and polarization diversity, two extremely effective methods that are particularly well suited to active antenna technology.

What the Ionosphere is Like

I've been studying the ionosphere, both as a hobby and as a profession, for most of my adult life, and it still amazes me that ionospheric propagation works at all. In fact, the more you know about the actual physics involved, the more you realize how unlikely it is that all the necessary ingredients can come together in just the right proportions to make radio propagation happen.

Perhaps the most astonishing thing about the ionosphere is just how little stuff is actually up there. The particle density of the ionosphere at altitudes that are interesting and useful, radio-wise, is on the order of *one quadrillionth* of the density at sea level. This is because the Earth's atmosphere is *compressible*, which means the density drops off exponentially with altitude.

The active ingredients in the ionosphere, that is, the things that make radio "reflection" work are free electrons. Note that "reflection" is in quotes, because, while the net *effect* appears somewhat like reflection, the actual process is nothing so simple.

Now, while free electrons are *necessary* for radio reflection, their mere presence is by no means sufficient. Electrons in random blobs or clouds, no matter how vast the quantity, are of little use for radio propagation, except, perhaps in *sporadic E* propagation, which is actually rather anomalous. For "normal" HF propagation to occur, the free electrons need to be in smooth *layers*. If you know the first thing about electrons, you know that they *repel* each other, so large mobs of electrons are *not* inclined to line up in nice rows. Their *natural* inclination is to go wandering off into space, to get as far away from each other as possible. The only thing that keeps these electrons in any kind of order is the layer of positively charged *ions* from which they were torn in the first place. But what keeps the ions in order? Gravity.

So the complete picture is, approximately this. We have a neutral atmosphere held near the Earth by gravity. UV rays from the Sun slap a certain number of electrons away from the neutral molecules, creating ions. The ions are heavy enough to be held by gravity, but the electrons are not. However the electric attraction between the ions and electrons *is* enough to keep them in order. On top of this, the ionosphere is actually part of the *atmosphere*, which has, not too surprisingly, *weather patterns*. Is it any wonder whatsoever that the ionosphere can be a bit temperamental, with wrinkles, swirls, and holes? The fact that it ever works is nigh unto miraculous.

Selective

Another outstanding feature of the ionosphere is that it is *dispersive*. Dispersion is what causes a prism to separate light into different wavelengths by angle of refraction. The ionosphere does the same thing to HF radio waves, though unlike a glass prism, it's a lot "squishier." Because the ionosphere is constantly wobbling around, due to both atmospheric turbulence and random solar influences, any signal refracted from the ionosphere is going to wobble around as well. However, because the ionosphere is *dispersive*, as well as mobile, the refracted signals wobble around in frequency dependent ways. The net result of this is what is known as *selective fading*, where any signal that is not infinitely narrow, is going to be selectively refracted, by our squishy prism. The wider the signal, the greater the selective fading. Part of the incentive for developing single sideband was to reduce selective fading by reducing the overall channel width.

Perhaps nowhere is selective fading more pronounced…and destructive…as in frequency shift keyed signals, such as RTTY. The mark and space frequencies can fade at entirely different rates, even though they're only 170 Hertz apart. In fact, the mark frequency can arrive back on Earth at an entirely different location than the space frequency!

As early as the World War II, it was determined that RTTY reception could be dramatically improved by means of *spatial diversity*. Two widely separated receiving antennas, each feeding a separate receiver, would assure that as the signal wobbled around, at least one of the receivers would receive either the mark or the space frequency. If the outputs of the receivers were combined, most of the signal could be recovered, even with severe selective fading.

Random Antenna Cure for Random Fading

One of the advantages of random wire antennas (especially long rambling ones) that we didn't previously discuss in that chapter, is that such beasties tend to have built-in spatial diversity…and sometimes polarization diversity. An antenna that covers a lot of real estate, even haphazardly, has a better chance of intercepting most of a wobbling signal than a more localized antenna. Of course, this concept can be taken too far, as long random wires can also have deep nulls, which *may* be right where you don't want them.

Intentional

That being said, a more intentional and deliberate approach to selective fading remediation is preferable. In general, selective fading is more pronounced along the basic direction of travel. In other words, it is best to have one antenna closer to the signal source than the other one, rather than both antennas the same distance from the source, though widely separated. This of course, presupposes that you have some idea which direction the signal is arriving from.

Traditionally, spatial diversity setups used identical receivers with the detector stages coupled together, so that the AGC action of both of them tracked. A few top-end military boat anchors, such as the Collins R390, were set up for diversity reception, and worked admirably.

Beginning in the 1930s there was a flurry of articles about diversity reception, but it was relatively difficult to implement. *QST* articles from May 1936, December 1937, April 1941, and April 1966 discussed the method in detail. However, most of these methods were rather difficult to implement, since, ideally, all the local oscillators of the different receivers had to be phase locked, or at least on identical frequencies. Since the trend had always been toward *more* conversion stages, it was somewhat of a nightmare to properly realize diversity reception.

This is *not* the case with the modern direct conversion (DC) receiver, where you only need one stable local oscillator, which can drive the mixers in any number of direct conversion receivers. You can then combine the audio outputs of these receivers for a true spatial diversity system.

We should note here that there are a number of methods that are referred to as diversity methods that are very different from the traditional "spatial" diversity understanding. *Frequency diversity* is often used in microwave communications systems as a means of reducing multipath and other types of fading. *Time diversity* can also be used. This is the repetition of a message over the same channel which may be subject to fading.

Modern signal processing and digital modes give us a lot more capabilities for this type of diversity. However, for this chapter, we have primarily focused on strictly *antenna diversity* methods, which implies either spatial or polarization diversity. The DC receiver may make experimentation with spatial diversity practical, at last.

The Case for Lots of Antennas

While there are plenty of examples of well-designed diversity antennas in commercial and military applications, true spatial diversity is used in relatively few amateur stations. However, most hams of long standing know intuitively that two antennas are better than one, and many antennas are even better than two.

At the very least, I strongly recommend that every radio amateur have at least two receiving antennas, one horizontal and one vertical. These can, naturally, be part of one's transmitting arsenal. Neither one of these has to be a great antenna, but they should both exist in one form or another. Having both a vertical and horizontal antenna at your disposal usually gives you at least two types of diversity...and sometimes three. The most obvious diversity is polarization diversity; with both a vertical and a horizontal antenna, you can *usually* receive a signal of either horizontal or vertical polarization. But the advantages don't end there.

A horizontal dipole at practical heights above real ground is unlikely to have any sharp nulls in the azimuth...but you can never be sure. A vertical antenna, on the other hand, is the only way to achieve true omni-directionality. It can be a great thing to have if you aren't *sure* that weak signal from the northwest is just a weak signal...or it's in your dipole's pattern null. Conversely, a vertical antenna, while being indiscriminate in the azimuth *may* have a sharp null in the elevation — especially when used for multiple bands, as many receiving antennas are.

A vertical antenna will, generally speaking, have good low-angle response. A horizontal antenna, at low heights, is preferable (or sometimes absolutely necessary) for high angle-of-arrival signals, such as NVIS signals. Having both vertical and horizontal antennas gives you a high degree of *directional* diversity.

Finally, by situating a vertical and horizontal antenna some distance from each other, you can achieve a high degree of *spatial* diversity, which is what you like to have for selective fading reduction — and sometimes even "normal" fading reduction.

It can be truly enlightening (and fascinating) to listen to a distant station such as WWV through two different receivers and antennas, one vertical and one horizontal. The fading (which will almost certainly exist to some degree over a long path) will nearly always occur

How Diverse?

One of the questions that was never fully answered, is, for an HF ionospheric communication system, just how far apart should the antennas in a diversity system be? And how many antennas and receivers do you really need? The simple answer to question one is, as far apart as possible. And the simplest answer to question two would be, as many as humanly possible. A less flippant answer would be: a minimum of a few wavelengths, and probably three antennas and receivers.

One army training manual from the post World War II era suggests placing three antennas in an equilateral triangle about 300 feet apart. However, there are so many variables, as we discussed in our introductory

---

at quite different times in the two different receivers. This in itself dramatically demonstrates just how flexible and "wobbly" the ionosphere is, even in the most stable of times.

A Different Slant on the Matter

There are times when even the most dedicated ham (present party excluded) has an aversion to paving the Back Forty with copper wire and filling the sky above said environs with more of the same. For those unfortunate souls, there is a viable option for obtaining at least *polarization* diversity with just one unobtrusive wire. It is the *slanting* wire. A wire antenna, either resonant or not, tilted at about a 45° angle with respect to vertical, will respond to both horizontally and vertically polarized signals.

Curiously enough such an antenna over real earth also exhibits a degree of *circular* polarization (or at least *elliptical* polarization) in some directions. This can be put to some use for investigating X and O modes of propagation.

See the antenna pattern in **Figure 20.A**. The lightly shaded "hemisphere" is the direction in which the radiation is right-hand circularly polarized, while the darker shaded "hemisphere" is the direction in which the signal is left-hand circularly polarized. Note the *degree* of circularity is not the same throughout the hemisphere in question. In a *NEC* output file, the parameter listed as *axial ratio* will tell you the degree of ellipticity of the CPOL signal. However, the amount of available circularity is *significant* throughout the region. The only direction in which true linear polarization exists is in the white sliver in the plane of the triangle formed by the tilted wire and the earth.

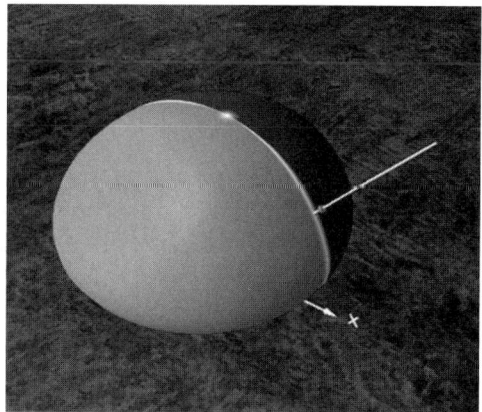

Figure 20.A — Modeled 3D antenna pattern of a slanted wire antenna over real earth.

paragraphs, that the only way to know for sure is with experimentation. Not every ham has the real estate to distribute three full-sized 80 meter dipoles around a 300-foot-leg triangle.

In some cases it might be relatively convenient to set up a few very small *active* antennas in a diversity configuration. In Chapter 22, we will describe just such an instrument, a three-way direct conversion receiver diversity setup, which should allow convenient experimentation. And it may actually prove to be practical for reliable communications.

Why Three?

As mentioned previously, selective fading occurs along the path of travel, so it's best to have one antenna "in front" of the other by several wavelengths in order to mitigate fading. But if you don't know the direction the signal is coming from, a triangular array is best. No matter which direction a signal is coming from, one or two antennas will be closer than the third. So, in effect, this arrangement is fairly omnidirectional, while potentially offering excellent diversity reception.

Some More Polarizing Thoughts

Often, the polarization of an incoming HF signal is unknown or highly variable, for much the same reason that selective fading occurs: the ionosphere is mobile!

That being said, there's a lot of erroneous literature in ham circles (and commercial ones, as well) that suggests that HF radio signals are *randomly* polarized. This is not the case. Ionospherically refracted signals are either X-mode or O-mode, (circularly polarized) as described in Chapter 15.

Now, the polarization may *appear* to be random, because, very few hams are using circularly polarized antennas. A CPOL (circularly polarized) signal of either sense can be received by any linearly polarized antenna, which would certainly lead one to believe that the polarization is random, without further information.

In addition, even if you *are* using a CPOL antenna, it will not "decode" the polarization unless it's properly bore-sighted. A CPOL turnstile antenna is only circularly polarized for signals arriving perpendicular to the plane of the antenna. The turnstile is linearly polarized to signals arriving from the edge (parallel to the plane of the antenna).

The ideal way to deal with circularly polarized, ionospherically refracted signals is with a properly oriented CPOL antenna of the proper sense. However, since, as of this date, an insignificant number of hams are using CPOL antennas, one way of dealing with the apparent random polarization (which manifests itself as fading) is by means of polarization diversity, which may be used separately, or in conjunction with spatial diversity.

The simplest way to achieve polarization diversity is with two antennas, one being vertical, and one being horizontal, each feeding a separate receiver. However, depending on the vertical angle of arrival, this can result in a very asymmetrical situation. Most practical dipoles at reasonable heights are fairly omnidirectional, with no significant "cone of silence" off the ends. However, a *vertical* antenna, either in the form of a vertical dipole or a grounded monopole, most *definitely* has a cone of silence directly overhead. So connecting a fairly normal vertical and horizontal antenna in a diversity arrangement will tend to be rather asymmetrical. This is not necessarily a problem, but you need to be aware of it.

Incoming!

For signals arriving at low angles, essentially parallel to the surface of the Earth, horizontal or vertical polarization are easily defined. But how do you define the polarization of a signal coming straight down from the heavens? If the direction of travel (Poynting vector) is "straight down," then the polarization *must* be parallel to the ground. It is horizontal, but there is an ambiguity of its azimuthal orientation. Vertical and horizontal lose their normal definitions. In this case, it's more meaningful to describe the polarization in terms of a North-South (N-S) axis, and an East-West (E-W) axis. A pair of diversity connected receivers, one connected to an E-W dipole, and one connected to an N-S dipole generally make a fairly symmetrical and effective polarization diversity setup, especially for NVIS signals. Which is a nice lead-in to our next chapter, on NVIS receiving antennas.

Chapter 21

NVIS Receiving Antenna

I wish that I had written this book during the peak of a sunspot cycle. Things would have been a lot simpler. When propagation is "hot," just about any random piece of wire will suffice for receiving countless hordes of super powerful signals. But then again, I wouldn't have much to write about.

It is when conditions are less than ideal that radio amateurs can show their ingenuity and practical knowledge. As I write this, we are on the downward slope of Solar Cycle 24, and it's been a weird ride for the past 11 years. On the average, it's been one of the lamest cycles in recent history, but it's also had some inexplicable, though brief, spasms of *inspiration* during its last gasps…or *something*.

That being said, for the next few years, propagation on the high bands will be in decline, which means we will need to be concentrating our efforts on the low bands. There may be some bursts of high-band sporadic E or other unusual propagation modes throughout the remainder of the propagation valley, but the overall trend will be downward in frequency.

Something Old, Something New

NVIS stands for *near vertical incident skywave*. Like so many buzzwords promiscuously cast around the amateur fraternity, NVIS is often treated as if it's some new invention. It's neither new, nor is it an invention. It's a technique. For many decades, NVIS didn't even have a name, even though it was commonly used and considered just part of normal HF propagation. If there's anything new about NVIS, it is a greater awareness of the importance and usefulness of this mode of operation, especially as the solar cycle declines.

Fog Nozzle

It's probably safe to say that most hams have no idea how far the radiation pattern of their antenna departs from the theoretical. Many hams talk about their super-duper directional antennas as if they're some kind of laser beam, when in reality, even under the best of conditions, the pattern is more like that of a fog nozzle. That's the thing firefighters use to cover as much volume of interest as possible with the least amount of water, with only a vague semblance of directionality.

Any real-world horizontal transmitting antenna is going to emit a combination of low angle and high angle radiation, in some proportion. (The same applies to receiving antennas, of course.) Even a three or four element 20-meter monobander at 60 feet will emit some degree of NVIS propagation. The only antenna that *won't* have any NVIS propagation is a *vertical* antenna, which has a very real and a very deep (though admittedly narrow) "cone of silence" in any direction suitable for NVIS propagation.

So it is not a stretch to say that any horizontal antenna at elevations significantly less than a half wave is an NVIS antenna...to some degree. However, a "real" NVIS antenna is one in which *most* of the radiation is directed at very high angles. Again, this can be said for a receiving antenna. This is one area where reciprocity rules.

Whys and Wherefores

Before we go too far into the mechanics, it's fair to ask why one wants to use NVIS propagation in the first place. NVIS is extremely useful for *reliable* low-band communication within about a 200 to 400 mile radius. For us KL7s, this means that most of the state of Alaska can be reasonably covered reliably with NVIS methods — and indeed this is usually how we do it. It's not an exciting mode for DXing, but it is an extremely *useful* mode for public service and emergency communications. When properly implemented, NVIS can provide very strong, stable signals within a region. Incidentally, our relatively new 60 meter band is ideally suited for NVIS communications in Interior Alaska, as we've discovered. The whole idea of NVIS is that you send a signal at high vertical angles and receive a signal that's coming nearly straight down. A few conditions need to be met for this to happen, of course.

1) You have to transmit a certain amount of energy at high angles. Again, this mostly precludes a vertical radiator.

2) You must operate *below* the critical frequency of the ionosphere. Frequencies *above* the critical frequency go right off into outer space. Since the critical frequency decreases after local nightfall, NVIS is *usually* considered a late afternoon or early evening mode. Typically, you want to operate about 10% below the critical frequency for optimum results.

3) You must operate *above* the D-layer absorption frequency. This generally precludes daytime 160 meter operation, and almost certainly precludes operation on 630 or 2200 meters at any time.

Sweet Spot

Although this can be highly variable, NVIS is most commonly and effectively used between 80 and 40 meters. In Interior Alaska, the critical frequency is seldom above 7 MHz, and usually considerably less. In mid-latitudes, the critical frequency can be as high as 10 MHz during solar maxima. Most of the time, it hovers in the 7 MHz region during daylight hours. When determining if NVIS is going to be practical for you, your local ionosonde is your friend. There is almost certainly a Digisonde near enough to you to be useful. For more information, see **www.digisonde.com/stationlist.html**.

Loops and Hoops

A one-wavelength horizontal loop elevated not more than 1/4 wavelength above ground is one of the most effective NVIS antennas. Many hams in rural Alaska have the space…and the need…for such an antenna. In this configuration, it acts similar to a cubical quad standing on end. Some hams lay a reflecting loop on the ground, to further emulate the cubical quad mode.

As we have discussed from many different angles, an active receiving antenna of just about any configuration can be made considerably smaller that its transmitting counterpart. The NVIS receiving loop is no exception. However, an *array* of small NVIS receiving loops is even better. With an array, you can build a much more directive antenna than you could with a single active loop. But beyond that, with an array of loops, such as in a four-square configuration, you can fairly accurately measure the vertical angle of arrival of an incoming signal.

Why would you care? Glad you asked!

Knowing the angle of arrival of ionospheric signals is an important part of understanding ionospheric physics. But, perhaps you're saying, *I don't care about the physics, I just want my radio to work.* To which we would say, *Why not do both?* NVIS transmission and reception is central to most ionospheric studies. If you don't believe this, take another look at the HAARP array…180 circularly polarized NVIS turnstile antennas (360 turnstiles in all, if you count the upper and lower elements separately).

Doing Pennants

We have already described such intriguing terminated loops as the Pennant, Flag, and EWE antenna. Now imagine standing a well-designed Pennant on end for a highly directive NVIS antenna. Or better yet, several on-end Pennant antennas. There are all kinds of exciting possibilities, since these Pennants can be *small*.

While most of these aforementioned antennas are usually oriented for vertical polarization, they are not strictly limited to that use. A Pennant antenna has a lot of performance for its size. You can turn it on end without too much mechanical misery; it lends itself to support by a single moderately sized mast. In fact, some ionosondes use a modified "upright" Pennant for broadband NVIS operation with a fairly controlled pattern.

Reflect on This

It's reasonably safe to say that with any horizontal antenna over typical ground, a good portion of the radiated power will be wasted. Take a simple dipole at *any* elevation above ground. Initially, half of its power will be directed below the horizon. Barring any ground reflections, none of that "downward" energy is going to do anyone much good. Fortunately, real ground (even a real *lousy* real ground) will contribute to the overall radiated signal. The actual character of your local ground will determine how much "downward-looking" energy is dedicated to heating earthworms, and how much contributes to useful reflection (and hopefully *reinforcement* of the overall signal.

Again, it's important to not underestimate just how much the conditions of the earth *well below* the surface of the earth can affect HF signals. In most cases, the fact that the true "reflecting plane" is well below the surface is a good thing — it makes your antenna appear higher than it really is. So, even if you incur losses by some degree of earthworm heating, the overall contribution of the earth is generally on the positive side of the ledger.

The complete truth is that the effort you put into creating a "perfect" ground plane may or may not be worth the effort. And in some cases, it might have a detrimental effect.

Most hams who *do* put any serious effort into ground systems, do so for *vertical* antennas, where the effort is not just nice, but absolutely necessary. For a horizontal antenna, especially one designed for NVIS purposes, an effective ground system can greatly improve performance, not only in terms of efficiency, but in the predictability of the radiation power. Which is nice.

One of the features of the HAARP array that is not clearly visible in most photographs is the extensive elevated ground screen. Acres and acres of aluminum clad steel wire in a grid structure, about fifteen feet above the tundra, comprise a near-perfect ground plane for the array. By the way, it may seem strange to use something like HAARP, a facility that radiates more HF power than any other radio station on earth, as an example for *receiving* antennas, where the radiated power is essentially zero. However, despite the vast differences, the principles of HAARP serve as a very useful benchmark, an ideal situation, especially for NVIS propagation. It's a very distant target to aim at, but if you at least start in that direction, you'll have a better NVIS antenna (for either transmitting or receiving) than if you started out in any other direction!

Chickening Out

There are two common approaches for building an effective ground plane for a horizontal dipole (or any other horizontally polarized antenna), both of which are rather different from the normal vertical radial system.

One method is to use resonant reflectors underneath the antenna, or array of antennas. We touched upon the full wave loop, with a tuned reflecting loop directly underneath. Lots of Alaskan bush hams use this method. The problem with resonant ground reflectors like these, however, is that they work over only a narrow range of frequencies, just like any resonant reflector in free space.

If you're using just a dipole, a tuned reflecting wire, either directly on the ground, or slightly elevated, is somewhat effective. Several parallel reflecting wires spaced a few feet apart are even more effective. But the problem is, multiple parallel reflecting wires tend to have lots of interaction, which can make the proper tuning a bit tricky. And, additionally, the more of these you have the narrower and touchier the tuning becomes. (We did some extensive testing on this matter for an "extra-curricular" project at HIPAS Observatory, with some really interesting…and frustrating…results. The lab notes make some interesting reading, but are probably "beyond the scope of this course." I'd be more than happy to share some war stories on this in private for any interested parties.)

So, if you are one of those hams who prefer to actually get on the air and operate, as opposed to endlessly tuning persnickety ground reflectors, we suggest taking the easy way out. Lay out a carpet of chicken wire underneath your NVIS antennas. This is essentially the approach HAARP took, in their construction of a large elevated grid. It is inherently broadbanded, and as far as anyone can tell, just as effective as a tuned reflector, though using a lot more wire. However, for Amateur

Radio purposes, a grid of chicken wire is cheap. And we actually used the stuff at HIPAS Observatory, where, unlike HAARP, we were also pretty cheap. Fortunately, the Alaskan soil is very dry and non-corrosive, so this chicken wire ground lasted a long time. As they say on that internet thing, your mileage may vary.

By the way, there is expanded copper mesh available, for a price. This is actually used extensively in AM broadcasting, in addition to the requisite ground radial system, especially if the antenna is of a very high feed point impedance. At KJNP, where I was chief engineer for a quarter century, we had two half-wave towers (420 feet each) on 1170 kHz. The feed point resistance of each tower was around 450 Ω, with about the same amount of inductive reactance to boot. We were operating a rather critical array with stringent requirements on relative phasing and power imposed upon us by the FCC. We had tremendous seasonal instability until we laid out a lot of copper mesh near the tower. It really made a huge difference. But it also cost a lot. Well, enough of my high power war stories for now. I promise.

The bottom line is, chicken wire works. And there's no tuning involved. Chicken wire has a long honorable history among radio amateurs, and even some commercial radio facilities.

NVIS DX

The very term NVIS DX probably seems to be a contradiction in terms. And indeed, in most places it is. NVIS normally dictates very short distances, as some very simple geometry should reveal. But not every place is normal. If you'll bear with me just *one* more time, I'll tell one final war story that is quite revealing.

At HIPAS Observatory, we had a monster upward-looking log-periodic curtain array (**Figure 21.1**). It had a central tower of about 150 feet, with four log periodic curtains suspended at their upper ends with cables, forming a large pyramidal structure. We had originally installed this monster for some testing purposes. Originally HAARP was going to be built with dozens of these things. HIPAS, being the predecessor to HAARP, was selected to be the guinea pig. As it turned out, this antenna, as impressive as it was, couldn't handle the anticipated power that HAARP was going to use. So, they decided on the turnstile array, but left the log periodic pyramid antenna up north for us to play with. Which I did.

We had a Harris mil-spec HF radio (can't remember the model number) and a Henry 2K linear amplifier, which we had been using for some plasma experiments. So out of curiosity, I hooked up the Harris and the Henry to this log periodic monster, just to see if I could work anyone

Figure 21.1 — The upward-looking log-periodic curtain array at HIPAS Observatory.

anywhere. The antenna was basically flat from about 2 to 20 MHz...and aimed very precisely...straight up. So I fired up the ad-libbed station on 20 meters sometime in the early afternoon, not expecting to get much response...despite us being in a pretty good part of a sunspot cycle. The antenna was just *wrong* for DX of any kind. But I thought I'd give it a shot, just for jollies. With a Henry 2K, I figured I could at least work some locals.

Lo and behold, I immediately found myself on the receiving end of a DX pileup... the biggest one I'd ever experienced...with stations roaring in from over the pole, Russia, Eastern Europe, Asia, and Malaysia. It didn't make any sense, but I figured I'd enjoy my 15 minutes of fame.

But why?

One of the biggest assumptions hams make is that the ionosphere is flat, even, and horizontal. This is sort of true in places that don't have a strong geomagnetic field. In Fairbanks, however, the north magnetic pole is about 17° off vertical, and the ionosphere in the region is tilted on the average of about 60°. An NVIS signal only *comes down* NVIS if the reflecting surface is horizontal. If the reflecting surface is *tilted*, an NVIS signal *becomes* a low angle signal after reflection. This is something you have to experience firsthand to believe.

Now, there was certainly a lot of non-reciprocal stuff happening too at this time. I never was able to fully sort out the details. But the bottom line is this: an NVIS antenna can be a very effective DX antenna under the right conditions. You probably aren't going to be able to duplicate our monster log-periodic pyramid, but perhaps you can with a nice NVIS array of more moderate antennas. And just such an array is described in the nextl chapter. Which makes this probably a good place to stop.

The Ionosonde, NVIS Supreme

While not directly related to receiving antennas, the ionosonde (ionospheric sounder) is an important application of NVIS methods and antennas. But, even more importantly, the ionosonde is of inestimable value for any user of the HF spectrum. In addition, since most books on transmitting antennas usually begin with a dissertation on propagation, there is some poetic justice in the fact that the tutorial section of this book on *receiving* antennas *concludes* with a word or two about propagation. A short study of the ionosonde will do a lot to tie some loose ends together.

First and foremost, an ionosonde is a radar, more specifically a *pulsed* radar. Short radio pulses are directed straight upward where they are reflected by the ionosphere and returned to earth to be received and detected by an appropriate HF receiver. The height of reflection can be determined by the time-of-flight of the transmitted pulse. However, unlike most radars which look for reflections from a "hard" target, the ionosphere is rather soft and squishy. But most importantly, the ionosphere is highly *frequency dependent.* The height of reflection depends on the radio frequency of the pulses. The ionosonde typically sweeps over a range of about 2 MHz to 10 MHz, depending on local conditions.

While ionosondes respond to all the ionospheric layers, the primary interest is in the F layer or layers. In the F layer, as the frequency is increased, the time of flight of the pulses also increases, until the point is reached where the pulses pass right on through the ionosphere. This frequency is known as the critical frequency, and the height at which it happens is the critical height. These figures are dependent on the time of day, and where we are in the sunspot cycle, among other factors.

While the critical frequency and height only apply to vertical incident skywave (VIS) signals, and is a *worst case* indication of propagation, we can derive the *maximum usable frequency*, or MUF from the critical frequency and height. While *oblique sounding ionosondes* also exist, which can measure MUF directly, they are far less common.

A particular implementation of the ionosonde, namely the Lowell Digisonde, dominates the ionosphere "business." A large worldwide network of Digisondes is available online: **www.digisonde.com/stationlist.html**. Lowell also has an excellent primer on ionospheric sounding on their website: **www.digisonde.com/instrument-description.html**.

While there is a lot of data that you can glean from an ionogram, and it can take a whole career to fully comprehend it all, there are a few simple morsels of information even a novice can grasp from the "picture" (**Figure 21.A**). Of most usefulness is the MUF, which is shown on a horizontal scale just below the ionogram. The MUF is given for several different distances (D). Naturally, these distances are for a single hop. The O and X modes are shown on the actual graph, in red and green, respectively. (Of course that doesn't show up in black and white in this book, but you'll see

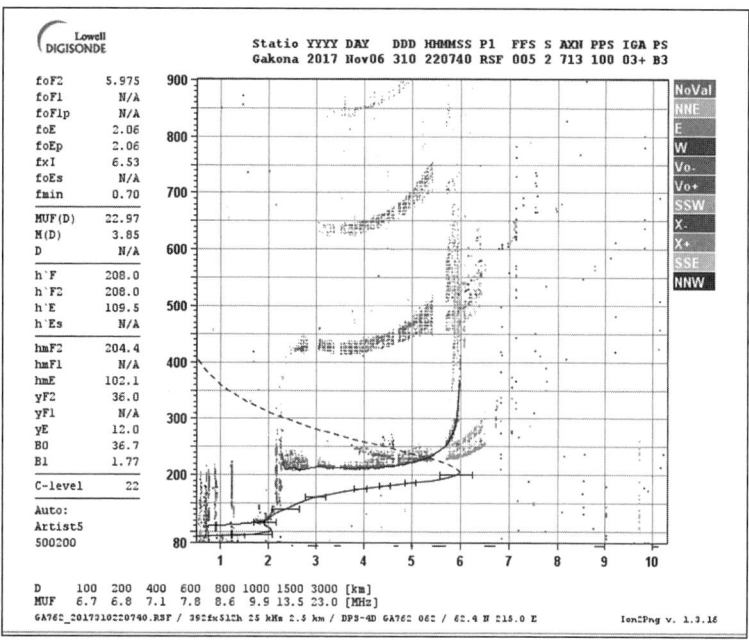

Figure 21.A — A Digisonde ionogram from HAARP.

the colors when looking at an actual ionogram.) The Digisonde also gives Doppler information, which shows how the ionosphere is moving. As the ionosphere advances (approaching local noon) the red will be shifted toward the blue. As it recedes, after nightfall, the red will be shifted toward the yellow.

When absorption is low, as in this particular case, you will see multiple reflections of the O mode signal…and sometimes a weak second reflection of the X mode wave (barely visible here). The black curve near the lower left is the *electron density profile* which shows the general concentration of free electrons, which pretty much determine how useful the ionosphere is for radio propagation. Maximum free electron concentration usually peaks in the 200 kilometer range.

Whether you're a short wave listener or a hard core DX chaser, understanding how to interpret ionograms will greatly increase your enjoyment of HF propagation. There's a lot of great science free for the taking, just by checking in with your nearest Digisonde. Take advantage of it!

Chapter 22

Receiving Antenna Projects and Accessories

Paul Graham, entrepreneur and author of *Hackers and Painters*, wrote a wonderful essay on engineering design. He says, among other things, that "good design is suggestive." What he means is that an intelligent person can take the principles of a well-designed widget and expand it to his or her own needs, improving on the original intent of the original designer.

It is in this spirit that we present this collection of receiving antenna projects. Rather than just offering a lot of cookbook designs for antennas, we instead will give you *frameworks* of receiving antennas and associated circuits that you can expand and adapt to your own particular situation. Just about any of the principles we have described so far can be combined in a myriad of antenna systems, simple or fancy.

Being a Model Citizen

Because antenna modeling programs have become so plentiful and cheap — and *free* is about as cheap as you can get — we encourage you to model some of these concepts to your own satisfaction before spending a lot of time and money on hardware and labor. But never forget that even the best computer modeling is no substitute for understanding and experience. Nobody appreciates the simplicity and ease of computer designed antennas more than someone (like me) who has spent countless hours in the field tweaking and pruning, and building, and rebuilding genuine copper, and steel, and bamboo, and PVC antennas, before such fabulous tools were available. But we old timers know the limitations too. And that can only come through practice.

The projects described in this chapter were designed specifically to give you a genuine *understanding* of the concepts we have discussed. We are fully committed to the following concept:

A Minute of Measurement Trumps a Decade of Debate

So, with that introduction, we trust that at least one of the following projects will pique your interest. Again, you may find that you can combine several of these projects into a system. Learn and enjoy!

Here's what's ahead in this chapter:

1) A simple and sensitive field strength meter.
2) A basic low band receiving loop.
3) An active helical frame loop antenna for 630 or 2200 meters.
4) A Trimpi Loop antenna.
5) The eXOgon antenna for exploring X and O propagation modes.
6) A four element NVIS Pennant array with Butler matrix.
7) An active ferrite loopstick for 630 meters. Great performance in a semi-tiny package.
8) A newfangled Adcock antenna: The Adcock antenna is a tried and true direction finding antenna, largely immune to skywave errors. Active methods enhance the Adcock even further.
9) An S-meter for direct conversion receivers.
10) A CPOL antenna diversity array using three antennas.

Project 1: KISS (Keep It Super Simple) Relative Field Strength Meter

This project has three goals in mind, the first being to demonstrate to the yet unconvinced that an extremely short antenna can actually receive something. And the second, to serve as an eminently useful tool around the shack. And finally, the third goal, to familiarize you with the extremely useful AD8067 chip. This is not an instrument grade device, by any means, but is sensitive enough to perform a number of useful tasks, such as verifying that a nearby transmitter is actually radiating.

Many hams, both new and old, tend to obsess over SWR and antenna matching, often investing hundreds of dollars in the latest and greatest analytical gadgets, without giving an iota of thought as to whether all their effort actually *accomplishes* anything meaningful in terms of radiating a signal. It is truly amazing how few hams actually have an instrument of *any* kind capable of giving any kind of meaningful feedback about their SWR obsession. It would seem to make sense to actually *measure* the effect that all one's knob twiddling has on the radiated signal strength, which again, should be the end result.

Fortunately, you don't need an NIST standard field strength meter to perform such a task; a rudimentary *relative* field strength meter is more than adequate, in the vast majority of cases.

In Passing

The passive field strength meter has been around for a long time, and at one time was part of the RF arsenal in most ham stations. The passive field strength meter typically consisted of a short whip antenna, an RF rectifier (or sometimes a voltage doubler), and a microammeter. These usually included a sensitivity adjusting potentiometer, as well.

There is nothing wrong with the passive field strength meter in principle, but they do tend to be a bit insensitive at the lower HF range of frequencies, which limits their usefulness in QRP situations. In addition, traditional microammeter movements are becoming rather expensive these days.

Better

An active field strength meter, on the other hand, can be quite sensitive. In addition, the use of an amplifier with a low output impedance allows the use of a lower impedance milliammeter, instead of the normal microammeter, which lowers the cost significantly. A more elegant revision also includes a logarithmic amplifier to expand the dynamic range, as well as eliminate the need for a sensitivity control in most cases. First we'll describe the one-chip version, which is about as simple as an active antenna can be. The KISS Active Field Strength Meter is shown in all its glory in **Figure 22.1**. Like all the projects described in this book, this is

Figure 22.1 — An active field strength meter.

only a suggestion. You can choose any form factor you like for the enclosure. Traditional passive field strength meters tended to be tall and skinny, but my station is laid out more horizontally, stylistically speaking.

The schematic for the KISS is shown in **Figure 22.2**, which you've probably noticed is similar to the one shown in Chapter 10, minus the following log-amp. The voltage doubling rectifier is shown in **Figure 22.3** and schematically in **Figure 22.4**. (By the way, this voltage doubling

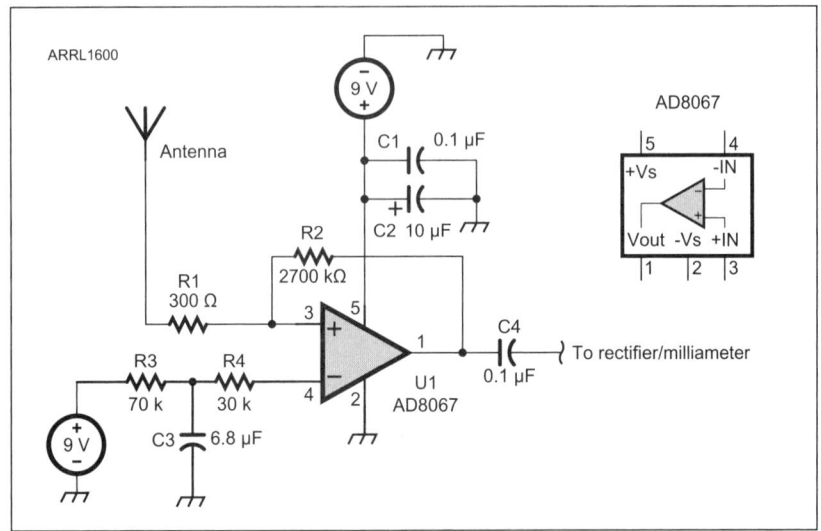

Figure 22.2 — At (A), the schematic diagram of the active antenna section of the field strength meter using the AD8067 op-amp in an inverting configuration. The photo at (B) shows the AD8067 (only available in a surface mount package) on a converter "daughter board" (the vertical PC board) which is then soldered to a larger PC board. We have used this circuit and construction method on numerous projects throughout this book.

Figure 22.3 — The voltage-doubling rectifier using a pair of 1N34 diodes.

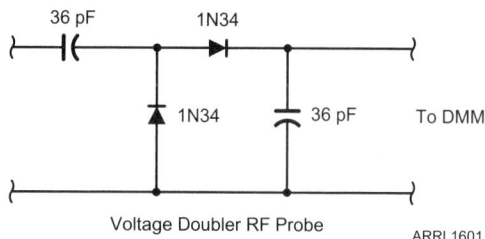

Figure 22.4 — Schematic of the voltage doubling rectifier circuit.

probe is a great thing to have around the ham shack, all by its lonesome. I use it frequently for troubleshooting and aligning receivers, when I don't have an oscilloscope handy.) A closer view is shown in **Figure 22.5**. (Yes, that is a shish kebab skewer used for the antenna. I have not performed a detailed *NEC* analysis of a shish kebab skewer, leaving that as an extracurricular project for the builder.) A collapsible whip would also work nicely in this application, and that has been used for a number of low-cost field strength meters over the years. If you like, you can add a log-amp (**Figure 22.6**) to your field strength meter. These useful devices are discussed in more detail in Chapter 10.

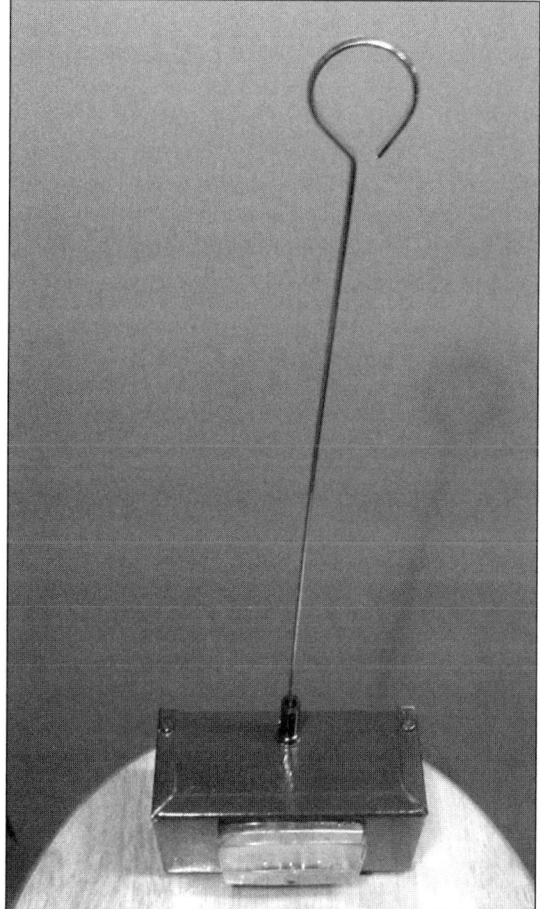

Figure 22.5 — The field strength meter uses a common shish kebab skewer as the antenna.

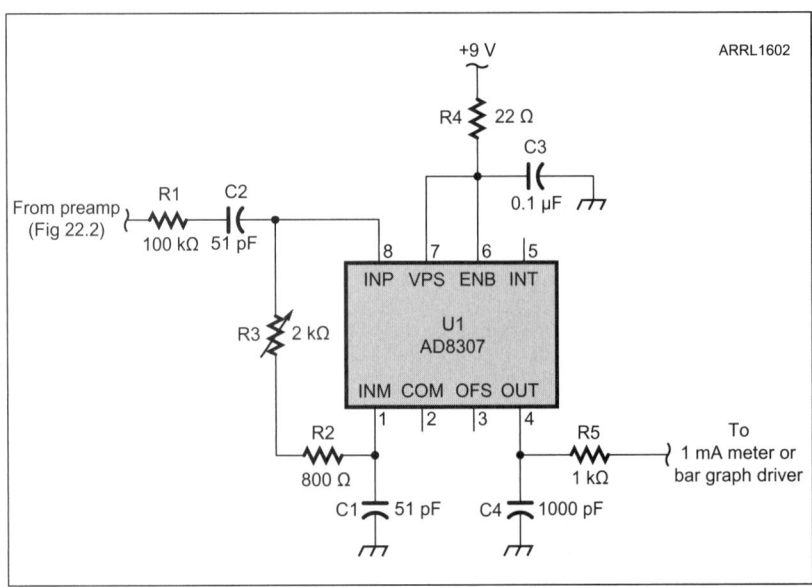

Figure 22.6 — Ambitious builders can add a log-amp to their field strength meter.

Project 2: A Basic Low Band Receiving Loop

There is nothing new or high tech about this project, but it's a good way to familiarize yourself with the surprising capabilities of "negative gain" antennas. By negative gain, we mean antennas with less gain than a "normal" dipole or ground plane antenna. This is a passive receiving antenna that is easy to adapt to any low band, but probably shows the most graphic results on 160 meters.

Tale of a Happy Ham

Alaska State Trooper Silas Hessler, KL3ZJ (**Figure 22.7**) lives about three miles from me as the raven flies, an ideal distance with which to test out various antennas under relatively controlled conditions. Until a year ago, Silas had never operated on 160 meters, and I "prevailed sorely" upon him to get on Top Band because I needed him to help me evaluate a new transmitting antenna I'd just erected (an inverted L). Alas, Silas

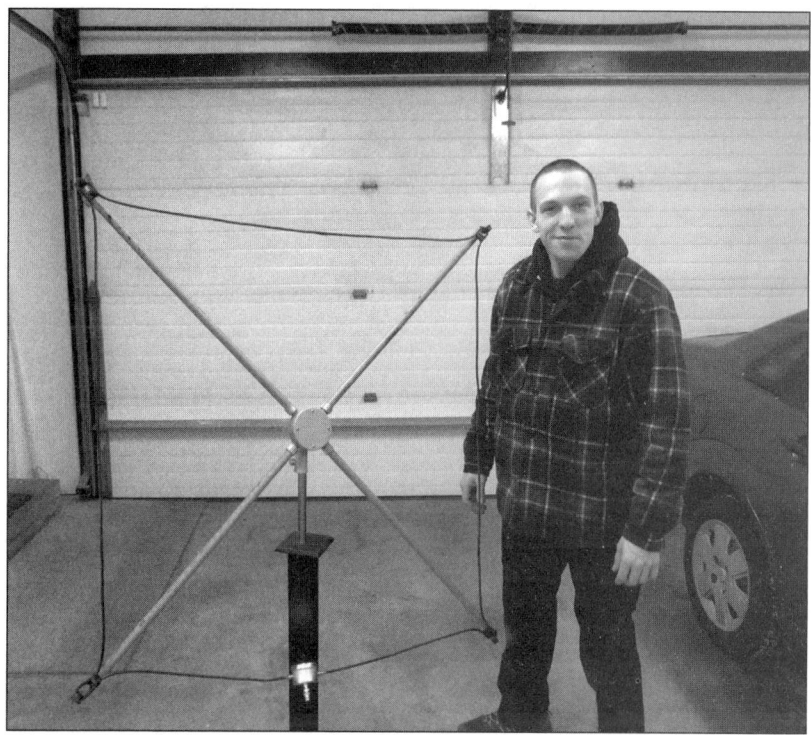

Figure 22.7 — Silas Hessler, KL3ZJ, with his basic receiving loop.

discovered his location had a nearby pole transformer that was blessing him with S-9+ noise 24 hours a day on 160 meters. It hadn't created any noticeable noise on other bands. Power line noise can be like that. The upshot of this was that Silas was *barely* capable of copying me on 160 despite the fact that he was basically in my back yard. Our friendly local power company (which, as a rule, has been *very* ham-friendly and cooperative) already had their hands full with more pressing projects, and informed us it would be a while before they could resolve the situation. In the meantime, I took this opportunity to demonstrate to Silas the wonders of small, shielded, single-turn, tuned loops with respect to local noise reduction. (And to remind myself of this fact!)

Taking a cue from a much-published similar project in the *ARRL Antenna Book*, we decided to toss together a smaller version and ended up with the loop shown in Figure 22.7. I already had several turnstile frames constructed for my eXOgon antennas, so I decided to re-purpose one of these for a shielded loop. I had thousands of feet of surplus RG-59 coaxial cable in the shack, just waiting for such a project.

Figure 22.8 — The loop shield is securely grounded to the junction box/feed point, and 1000 pF of capacitance is inserted between the two center conductors.

A small gap is cut in the shield of the coax cable at the top of the loop, opposite the feed point. The shield is securely grounded to the junction box/feed point, constructed of a small tea tin (**Figure 22.8**). Then 1000 pF of capacitance in inserted between the shield and the center conductor, which is in parallel with the internal capacitance of the coax loop. This was determined experimentally to broadly resonate the loop at 1.85 MHz. Because of the large C/L ratio of this loop, it is relatively low Q and works quite well across the entire bottom half of 160. Final tuning was performed using a signal generator and oscilloscope (**Figure 22.9**) to look for the greatest voltage across the loop. A dip oscillator or antenna analyzer could have been used as well.

Proof and Pudding

Frankly, I wasn't too sure what to expect from this antenna. I'd never had to contend with a noise level as severe as Silas's, but we figured there wasn't much to lose by trying it out. I instructed Silas to take it home, stick it in his lawn atop a small post and orient it broadside to the direction of the guilty transformer. Without getting too scientific about it, Silas

Figure 22.9 —Final tuning of the basic loop was performed using a signal generator and oscilloscope to look for the greatest voltage across the loop.

was shocked and awed at how he was able to almost totally null out the noise by appropriately rotating the loop. Considering that the construction of the loop was somewhat haphazard, I was a bit surprised myself. In any case, the improvement in signal-to-noise ratio was such that Silas and I could immediately hold high quality contacts on a regular basis.

Using a small "negative gain" loop such as this, will, of course, result in a significantly weaker signal than a full sized dipole or vertical, but the increase in intelligibility is well worth the loss of S-units. In fact, this is probably the *only* legitimate reason to use an RF preamp on any current model of HF receiver — in concert with a low gain antenna. The use of an RF preamp with a "normal" antenna is quite likely to decimate the dynamic range of your receiver, as described earlier in this book.

Incidentally, it would be a nice feature if, when switching in an RF preamplifier, the S-meter is compensated *downward*, so as to retain something like a valid signal strength reading (at least a relative one) regardless of the preamplifier setting. Some, but not many, current transceivers give accurate S-meter readings regardless of preamp settings. In any case, keep in mind that it is the signal-to-noise ratio that is the end goal, not wobbling the S-meter!

Project 3: An Active Helical Frame Loop for 630 or 2200 Meters

We were already in the throes of writing this book when we received the welcome news that at long last, we had *official* access to our two new Amateur Radio bands, one at 630 meters (472 – 479 kHz) and one at 2200 meters (135.7 – 137.8 kHz). We would be most remiss if we didn't include these wonderful new bands in this tome. The acquisition of these bands represents the culmination of 13 years of hard work by Frederick Raab, W1FR, Rudy Severns, N6LF, and others in the ARRL 600 Meter Experimental Group.

Frame loop antennas have been around as long as radio. Most early AM broadcast receivers used a frame loop antenna either sitting on top of the radio console, or nearby. These could be conveniently rotated for best reception. The better frame loops could occupy a fair amount of desk space, and while any self-respecting radio person thought they were things of beauty, said self-respecting radio persons often had to share living space with folks who were not *quite* as enamored with radio apparatus as themselves. So the generous-sized frame loop gradually slipped from the scene.

At least a couple vintage floor model AM broadcast band radios had rotatable frame loops inside the cabinet. As radios further shrunk in size, the conventional frame loop was replaced by a fairly miserable facsimile. A typical table radio had a frame loop wound around a wooden or cardboard panel which served as the back of the radio. This configuration was fairly standard for a couple of decades until the ferrite loop was developed. Actually the ferrite loop was a marked improvement over the back-panel frame loops, but could still be outdone by the larger "official" frame loops.

One interesting development of the frame loop in the early 1970s was not actually a new development at all, but a clever implementation of the much earlier *loose coupler*. It came under the name Select-A-Tenna, shown in **Figure 22.10**. It was a 12-inch-diameter plastic frame loop antenna with a 365 pF tuning capacitor in the middle. (These still show up online, but they're getting more difficult to find.) You would place the Select-A-Tenna next to your inexpensive transistor AM radio, with the loop more or less concentric with the ferrite loopstick in the receiver. This would greatly

Figure 22.10 — The Select-A-Tenna is a 12-inch-diameter plastic frame loop antenna with a 365 pF tuning capacitor in the middle.

increase the sensitivity of the receiver. When I was chief engineer at KJNP, we resold *hundreds* of Select-A-Tennas to listeners in the Alaskan bush. It was a very effective product, based on really good science.

I have modified Select-A-Tennas to use well below the AM broadcast band. And when I started working on the first draft of this project, I was going to describe just that. However, after learning that the original devices are not as plentiful as they once were, I've decided to describe an improved version that you can build from scratch. If you happen to encounter a Select-A-Antenna at a flea market, snatch it up. It makes a great platform for a number of low-band receiving antennas.

Spiral or Helix?

For the longest time, frame loop antennas were more pancake-like than cylindrical. Close winding a lot of turns gave you a lot of inductance, but also had a lot of interwinding capacitance, which can create some issues. A couple of methods were used to reduce interwinding capacitance. One of these was the *honeycomb* winding, where the coil was interlaced through slots in the form. The cardboard-backed frame loop table radios often used a honeycomb winding. What this configuration did was cause each turn to cross the adjacent turn at right angles, or as close to that as practical, reducing the capacitance significantly.

Another method that was commonly used was to space wind the coil in a spiral form. While reducing the capacitance severely, it also resulted in a much less efficient antenna, the outer part of the winding being much more effective than the inner ones. It also made calculating the thing much more difficult.

A third method is to space the windings in a *helical* fashion. This resulted in a more cylindrical antenna, which could occupy a lot less "sky" than the traditional frame loop, a small step toward miniaturization. The only real disadvantage of this is method is that the pattern is highly modified over the pancake-like version. It is far less effective for nulling interference or direction finding. If those factors are not a consideration, it makes an excellent antenna, and that is what I describe here.

This is a somewhat experimental antenna, and is certainly subject to a number of improvements. And it faithfully follows my prime directive, "First you make it work; then you make it pretty." While not an RF gangbuster, it will give you a sensitive antenna to get you receiving effectively on 630 meters with a minimal amount of real estate. (The transmitting antenna for this band is another issue.) I used a number of readily available components for this, so free feel to season to taste. When it comes to experimental (and occasionally permanent) antennas of any kind, I am an

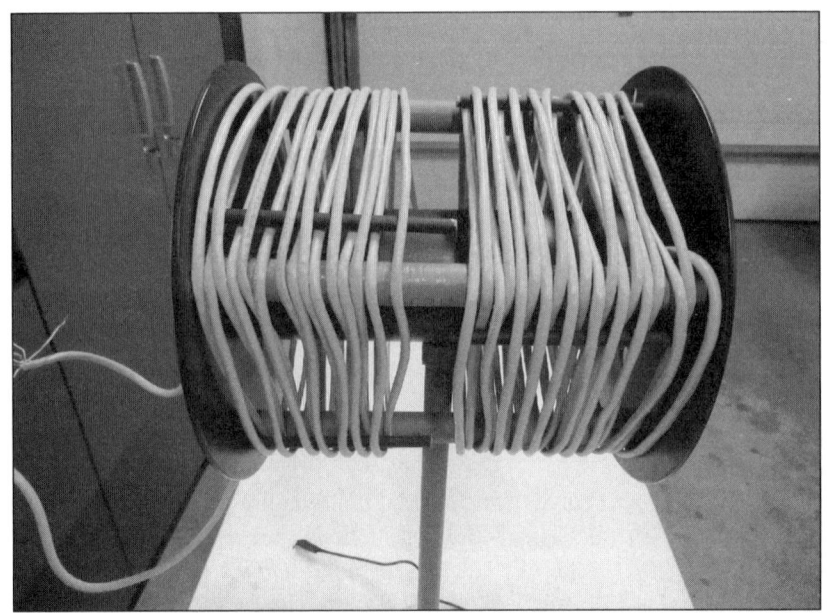

Figure 22.11 — At (A), the author's frame loop built around a large plastic wire spool, using 50 turns of CAT5 cable supported by four PVC rods. The photo at (B) shows an early version of the frame loop wound with no. 24 AWG magnet wire that proved difficult to space evenly.

inveterate fan of PVC plumbing. This includes not only the tubing, but the various connectors and T fittings available at the local home center.

I built my antenna around a large plastic wire spool. I installed four PVC rods between the opposite edges of the flanges, shown in **Figure 22.11A**. I used 50 turns of CAT5 cable, because I had a lot of the stuff lying around and it allowed me to fairly evenly space the turns without too much trouble. I made a first attempt with no. 24 AWG magnet wire, but keeping the windings evenly spaced was nearly impossible. **Figure 22.11B** shows my frowsy first attempt at this.

The nice thing about using something like CAT5 or other multiconductor cables is that you don't have to wind as many turns. In addition, the jacketing serves as a convenient space-keeper for the numerous windings. Simply "splice" the end of one conductor to the beginning of the next one, and so on. Since the conductors are color coded already, you don't have to worry about losing track of who's where. See **Figure 22.12**.

You can see that I also wound this in semi-honeycomb fashion… more to demonstrate the general principle than to truly take advantage of the method. I inserted small wooden dowels between the main PVC rods. There's already a lot of stray capacitance in multiconductor cables, which this particular honeycombing can't do much to fix, but if you are using individual wires, this is how to do it. Or at least one way to do it.

Notice how the cable is "woven" around the dowels. In fact, in this project, I actually use the stray capacitance to achieve self-resonance. With no external capacitance (other than my oscilloscope probe) I measured a Q of nearly 200 at 137 kHz (in the 2200 meter band). I did have to judiciously but aggressively use my diagonal cutters to achieve the final tuning, removing first a couple of turns and then a quarter of a turn to achieve final resonance, keeping in mind that I had to splice all the turns together again after each trimming. I'm sure there's a less tedious process available, but experimental projects normally rely on tedium.

My point in not using any tuning capacitors with this was to take full advantage

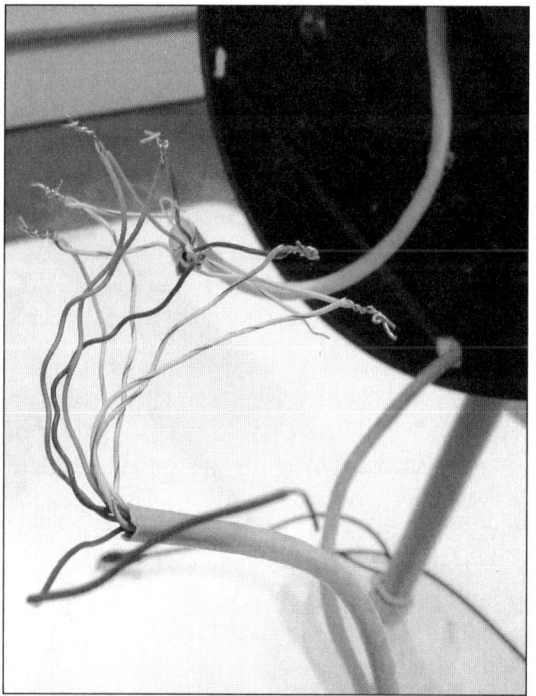

Figure 22.12 — The conductors inside the CAT5 cable are color-coded, making it easy to keep track of the wires as you are splicing them.

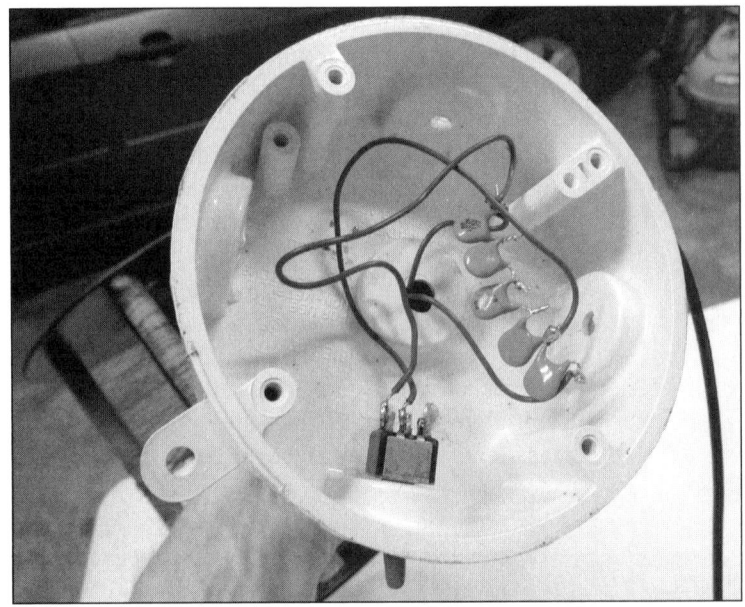

Figure 22.13 — Capacitors in base of helical frame loop pedestal can be switched in for two band operation. This particular loop has a fairly large L/C ratio so it covers either the entire 2200 meter or 630 meter band with ease. A sharper tuned circuit will benefit from a variable capacitor (or two) here.

of the extremely high input impedance of the AD8067 op-amp, which implies a very high L/C ratio. If you do need to tune this antenna, you can insert suitable switched capacitors, as I have shown in installed the base of the antenna (**Figure 22.13**), just as an example.

The output of this antenna goes directly into the input of the AD8067 in differential mode. If you're obsessive about balance, you can use *two* AD8067s in push pull, with an output transformer, as I have shown in the eXOgon antenna. This is probably overkill for this antenna, which doesn't have the best radiation pattern in the first place.

In Its Place

You can install the loop atop a short pedestal for convenient placement on your operating bench, which is how I operate mine. However, you may like this antenna so much you'll want to install it permanently outdoors, away from noise-generating house wiring and such. You can use an inexpensive security camera rotator to steer this antenna if you like. Like any other small loop, this also lends itself to incorporation in an *array* for even more effective beam steering.

We hope that if you haven't considered trying the "New Top Band," that you do so. This antenna will make your entry into the field simple and straightforward.

Project 4: A "Trimpi Loop" Antenna

We have quotes around the name of this antenna, because its developer, Dr. Mike Trimpi never officially named it as such. But he deserves full credit for this interesting antenna. The theory of this antenna has already been discussed thoroughly in Chapter 11, so we will concentrate here on how to get this up and running in the quickest way possible.

The Trimpi Loop, unlike the various directional terminated Flag, Pennant, and K9AY-like antennas is specifically *non-directional*. The priorities of radio science antennas are sometimes quite different from those of more conventional Amateur Radio antennas. The Trimpi Loop antenna should be considered more of a scientific instrument than a communications antenna. It was specifically designed for ionospheric research where very low noise, extremely flat bandwidth, and a very *symmetrical* pattern are foremost. For radio amateurs more interested in the radio science than "normal" amateur communications, this antenna shines.

As noted before, I'm a huge fan of PVC plumbing, and in the offbeat field of ionospheric research, readily deployable plumbing-based antennas are the norm. See **Figure 22.14**. A single 15-foot-long, 4-inch-diameter piece of PVC pipe is at the heart of the antenna. Like the K9AY Loop, there are two orthogonal loops suspended from the single central mast.

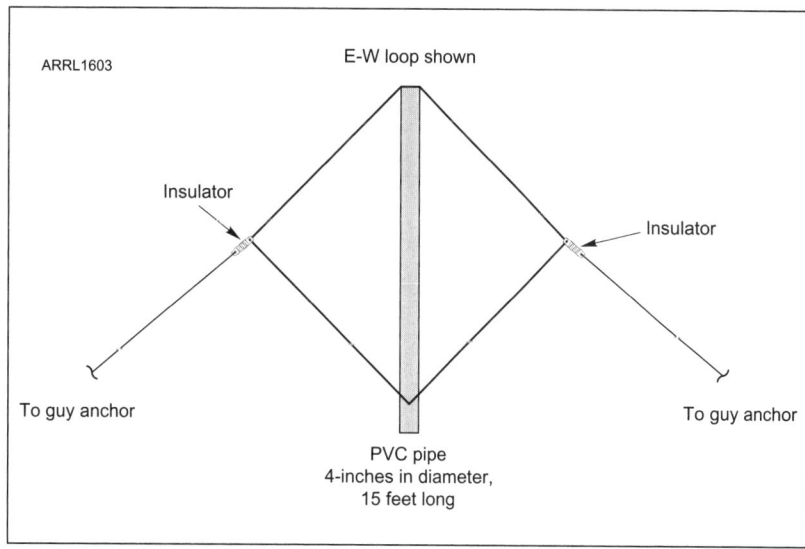

Figure 22.14 — A single 15-foot-long, 4-inch-diameter length of PVC pipe supports the Trimpi Loop antenna. Like the K9AY design, there are two orthogonal loops suspended from the single central mast, but only one loop is shown in this drawing.

However, the loops in original Trimpi antenna are not triangular, but square, standing on end.

The outputs from the loops are taken off the bottoms, making the antenna horizontally polarized and omnidirectional. In Figure 22.14, only the East-West loop is shown. The North-South is identical, but at right angles. Ropes or cords running from the insulators to the guy anchors serve as both mast guys and hold the shape of the loops.

The loops of this antenna are normally significantly smaller than a full wave, so as to not show any resonance in the frequency range of interest. Remember, a primary criterion is that the antenna response is extremely flat; the lossless feedback networks of the associated amplifiers maintain the flat gain of the antenna. The outputs of the antenna loops are combined in a broadband 90° hybrid network after amplification to achieve the omnidirectional pattern. This is the same method used in the eXOgon antenna, except that X and O switching is not normally necessary. It can be implemented, if desired, however.

The amplifiers can be installed in a small insulated box near the base of the antenna. Either battery or bias T methods can be used to power the amplifiers, as is the case with just about any active antenna.

The Trimpi Loop is an excellent antenna to use as a gain reference. It is a great antenna to use for general shortwave reception as well.

Project 5: The eXOgon Antenna for Exploring X and O Modes

This is actually the antenna that instigated the very writing of this book. I tell the story in detail in Chapter 15, along with a lot of the practical construction information, but I'll add a few notes here as well. This is one of the few examples in my ham radio experience that *Necessity* indeed was the *Mother of Invention*. I've always contended, and have written an article or two on said contention, that this proverb is fundamentally a lie. Most scientific progress results not as a result of trying to fix a problem but rather the happy result of some experiment having gone seriously awry. This is especially true in Amateur Radio. Most of my greatest intellectual revelations have occurred soon after regaining consciousness.

That being said, the creation of a wideband, circularly polarized, and…most importantly… *conveniently deployed* antenna was found to be absolutely necessary in order to allow any significant number of hams to become aware of X and O propagation. Understanding X and O modes of propagation is fundamental to understanding the ionosphere — and just about any phenomenon related to HF propagation. The fact that 99% of all radio amateurs have been able to get by without even knowing the existence of X and O modes simply shows how wonderfully forgiving radio is!

Circular polarization is to HF antennas what binocular vision is to sight. Sure, you can get by with one eye, but your perception is limited. Being able to separate X and O mode rays will open up whole new horizons for the radio experimenter…and average ham, too.

Figure 22.15 shows all the ingredients for the eXOgon. As with

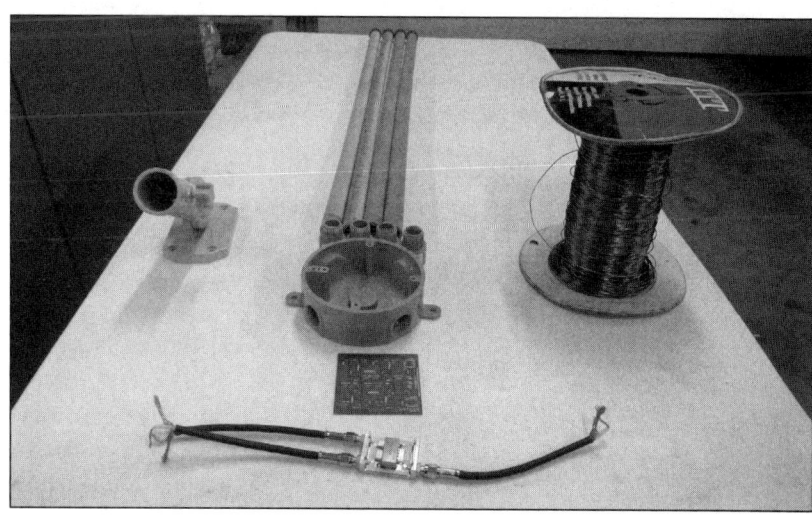

Figure 22.15 — All the ingredients for building an eXOgon antenna.

many of my other antennas, this design shows my infatuation with PVC plumbing. In this case, it's actually PVC electrical conduit, along with and associated junction boxes and connectors. There are a number of PVC-like plastics used in the plumbing and electrical industries, but as far as radio utility is concerned, they are all the same. I use the generic term PVC to include them all. Only at UHF frequencies might one conceivably detect any difference in the materials in applications such as this. At HF, we will be nowhere approaching conditions where the dielectric properties are meaningful.

The eXOgon is a small turnstile antenna, consisting of two crossed "arms" each being slightly over five feet long. The dimensions of the basic eXOgon were decided strictly on convenience; there is nothing meaningful "radio-wise" to the choice of dimensions. As it turns out, electrical conduit (either metal or plastic) comes in 10-foot sections, known in electrician vernacular as "sticks." With one stick I can make one eXOgon antenna. As it turns out the finished size is also very convenient mechanically.

The antenna conductors are four 2½ foot lengths of magnet wire, secured at the outer ends of the arms and terminating in the central hub, forming two crossed "dipoles." We use the term "dipoles" a bit advisedly. The truth is each "half dipole" forms an independent voltage probe; they don't really function as a dipole in the normal sense, as the impedance at the center is infinite. Both the North-South (N-S) dipole (referred to as a "channel 1"), and the East-West (E-W) dipole ("channel 2") and associated amplifiers are identical. There are four AD8067s in all, two for each channel. The outputs from the amplifiers are operated in push-pull, through a small trifilar transformer. As discussed in Chapter 15, where we show the circuit diagram for the amplifier, these transformers are entirely non-critical. Just be sure they're all wound the same, using the same type cores.

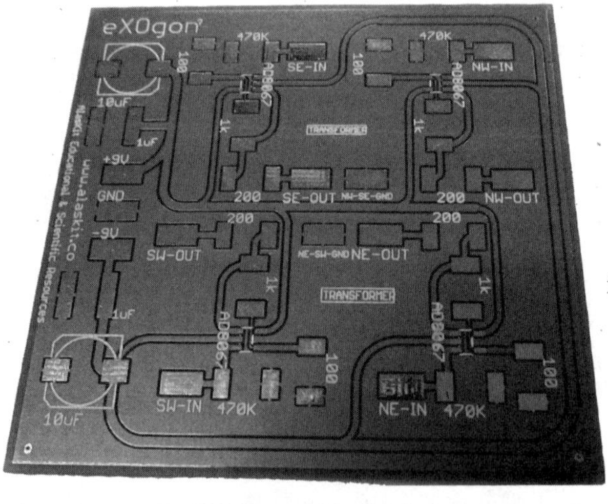

We have produced a small number of circuit boards (**Figure 22.16**) upon which the AD8067s and associated components can be soldered, available by request. We specifically designed the boards so that either surface mount or leaded components can be used. (This

Figure 22.16 — The AD8067s and associated components can be soldered to this circuit board.

was a gut wrenching decision, but it turned out to be the right one. A lot of hams, especially "properly seasoned" ones, are averse to surface-mount soldering. Except for the AD8067s, which are *only* available in surface mount packages, the rest is accessible to old folks and their soldering skills). You can also etch your own boards; we can supply the artwork for this as well.

We found that CAT5 cable works wonderfully for transporting the independent N-S and E-W channel signals to the shack. That also provides a means to power the amplifiers, through the remaining pairs.

Figure 22.17 — A finished eXOgon antenna.

The phase switching is performed at the shack end of the transmission lines. A small project box encloses the 90° hybrid and the phase reversal switch. We have successfully used 100 feet of CAT5, and there is no reason to expect that you can't use much longer runs, if necessary.

Although the characteristic impedance of CAT5 is about 100 Ω, there's no compelling reason to try to match the output of the amplifiers to this or to the input impedance of the hybrids. Although the 90° phase shift through the hybrids is specified with 50 Ω loads on all three ports, we have not been able to measure any additional phase error with 100 Ω source impedances. The Mini-Circuits hybrids are described and shown in Chapter 15, as well.

Once the turnstile and amplifiers are assembled, they can be mounted on a flag bracket, which is then mounted on a rod or post stuck into the ground. This allows you to swivel the thing up and down for elevation adjustment, and rotate it around the post for azimuth rotation. You can also mount this antenna on a small az/el rotator for convenient remote steering. **Figure 22.17** shows a finished eXOgon antenna.

When the antenna is "looking" horizontally, you want to be sure the arms form an X, not a +. This assures that the symmetry of both channels relative to the Earth is maintained. One way to calibrate the antenna is to use a vertically polarized radio source, perhaps 100 feet away, while "looking" directly at it with the eXOgon. In this condition, both channels should show equal amplitude and be in phase.

On the Air

The only way to fully understand the eXOgon antenna is to use it. Listen to a lot of distant HF signals, while switching between X and O. Observe the *profound* difference it makes on *most* HF signals. Once you've played around with this for a while, you will probably want to figure out how to build a circularly polarized *transmitting* antenna to explore the possibilities even further. But, that's the subject for another book.

Project 6: A Four Pennant Butler Steered NVIS Antenna

It's really fun when you can create a ham radio project that demonstrates several principles all at once. The genesis of this particular project was the Digisonde ionospheric sounder. While there are a number of Digisonde deployment schemes, the typical installation has a centrally located transmitting antenna (a broadband NVIS array), surrounded by three widely spaced receiving antennas. The purpose of the multiple receiving antennas is to determine the angle of arrival of high angle (NVIS) signals, or the horizontal drift of the ionosphere.

The typical Digisonde receiving antenna is an expensive beast, but it's small and compact. On the other hand, a typical Pennant antenna is larger...but cheap.

This project demonstrates the important concept of one-dimensional beam steering. While beam steering in two dimensions is possible, it does not yield itself to the simple Butler matrix. What we describe here is a four element, wide spaced, linear array with four different beam directions along a line of your choosing. Aligning your array with the *magnetic* north and south will give you the most versatility for radio science experiments, but you might find being able to scan east-to-west is most useful for typical operating.

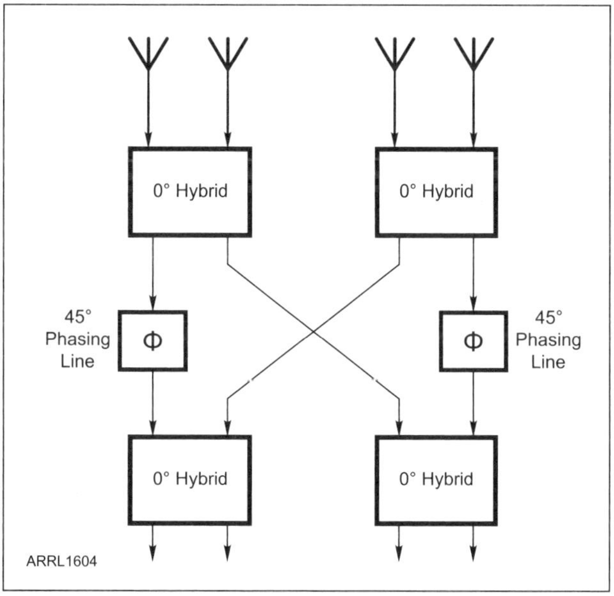

We've kept the details of this project very generic; you can tweak this to cover any frequency you like. The idea here is to use the well-known properties of a Pennant antenna, and stand it up on end. This gives you a broadbanded NVIS antenna. As with other "rapid deployment" antennas we've described, the "upstanding Pennant" can be erected with a single supporting mast, using the remaining antenna wire as part or all of a guy system.

Place the four Pennants in a straight line, evenly spaced and separated as widely as your available real estate or budget allows. A simplified four-channel Butler matrix is connected to each of the antennas with equal length transmission lines. See **Figure 22.18**.

Figure 22.18 — A simplified four-channel Butler matrix is connected to each of the antennas with equal length transmission lines.

Each of the outputs from the Butler matrix will be a different direction along the line of antennas. The actual angle will be dependent on the spacing of the antennas, which can be modeled accurately with a *NEC* antenna modeling programs. The beam can be rapidly steered by means of PIN diode switches. Mini-Circuits produces both active and passive PIN switches for applications such as this.

As you might suspect, this array is most useful for NVIS signals, but can be used for lower angle beam steering, with a progressively less effectiveness as the angle of arrival approaches the horizontal. Used in conjunction with conventional direction finding (azimuthal) methods, being able to derive vertical angle of arrival can help "DF" stations well beyond the horizon. So called "3D" direction finding is extremely useful for a lot of radio science experiments. This is just one more tool in your arsenal.

Project 7: An Active Shielded Ferrite Loopstick for 630 Meters

What's old is new again. Here is a classic shielded ferrite loopstick antenna with a modern edge.

One of the best things about active antennas is that the "signal grabbing" and impedance matching functions can be kept entirely separate. This is because the near-infinite impedance active devices used in all modern active antennas allows the antenna to operate in nearly ideal conditions.

Except for some rare circumstances, the antenna is connected to a receiver by some form of transmission line. Whether used for transmitting or receiving, the transmission line always has some effect on the antenna radiation pattern; it is nearly impossible to separate the antenna function from the transmission line function. While various forms of isolation in the form of baluns or chokes can largely reduce the transmission-line-to-antenna coupling, and deleterious effects, they are seldom complete. For absolute precise direction finding, the effect of the transmission line needs to be reduced to the vanishing point.

The use of a near infinite impedance op-amp, however, can reduce the coupling to the transmission line to a degree impossible to achieve with standard isolation methods. Of course, the use of an active device is only applicable to receiving antennas.

Shields Up

An electrostatic shield is often used on precision grade loopstick

Figure 22.19 — To maintain pattern integrity in the presence of nearly metallic objects, the loopstick is enclosed in an electrostatic shield — a U channel of aluminum.

antennas, for the same reason they are used on other types of loop antennas. That is, to maintain pattern integrity in the presence of nearly metallic objects. The loopstick is enclosed in a U channel of aluminum (**Figure 22.19**). It doesn't matter if the ferrite rod extends beyond the shield, what's important is that the actual windings are inside the shield. Half of a "Bud Box" cabinet serves very nicely as the shield. Another function of the shield in this project is to provide a convenient mount for the active preamplifier.

Drawing a Bead

If you've ever tried to acquire a large ferrite rod for constructing a loopstick antenna, you'll realize they aren't cheap. You'd think ferrite was made of Cold Pressed Latinum.

The last time I went to Dayton Hamvention, there was a vendor selling piles of large ferrite beads, the type you slip over coaxial cable to form a choke balun. I think I paid about $10 for a few hundred of them... not that I needed them at the time, but I know Cold Pressed Latinum when I see it.

Figure 22.20 — Ferrite beads about 1 long and ¾ inch in diameter, with a ¼ inch hole through the middle, are glued together to make the antenna core.

These beads (**Figure 22.20**) are about 1 inch long and ¾ inch diameter, with a ¼ inch hole through the middle. I decided to Super Glue about half a dozen of these end-to-end and see how they worked as a the core for a large loopstick antenna. I was greatly pleased.

Ferrite is non-conductive in the first place; each microscopic magnetic domain is completely insulated from its neighbor. This is why you can glue beads together to come up with one large rod. I happened to have in my vast archives *one* very large ferrite rod, which happened to be exactly the size of six of my hamfest beads placed end to end… minus the hole. Performing every test I could think of with my available equipment, I could find no difference in permeability or loss — despite the fact that the daisy chained beads had a quarter inch hole through the middle. So, I can with all confidence heartily recommend building what could be very expensive ferrite rods out of bargain basement beads. (If anyone has any further data or experience with this, I'm happy to hear of it.)

Using the empirical approach, I close-wound about 250 turns of no. 28 AWG magnet wire around the center of the rod as shown in **Figure 22.21**. There's really no point, other than esthetics, to wind the coil in nice neat layers. A "scramble wound" coil has just as much inductance as the "obsessive compulsive" winding, and *probably* has less inter-winding capacitance. Furthermore, while the more electrically pedantic in our midst will insist on winding such loopsticks with *Litz wire,* I have yet to see any glaring benefit to doing so. Like most of my receiving

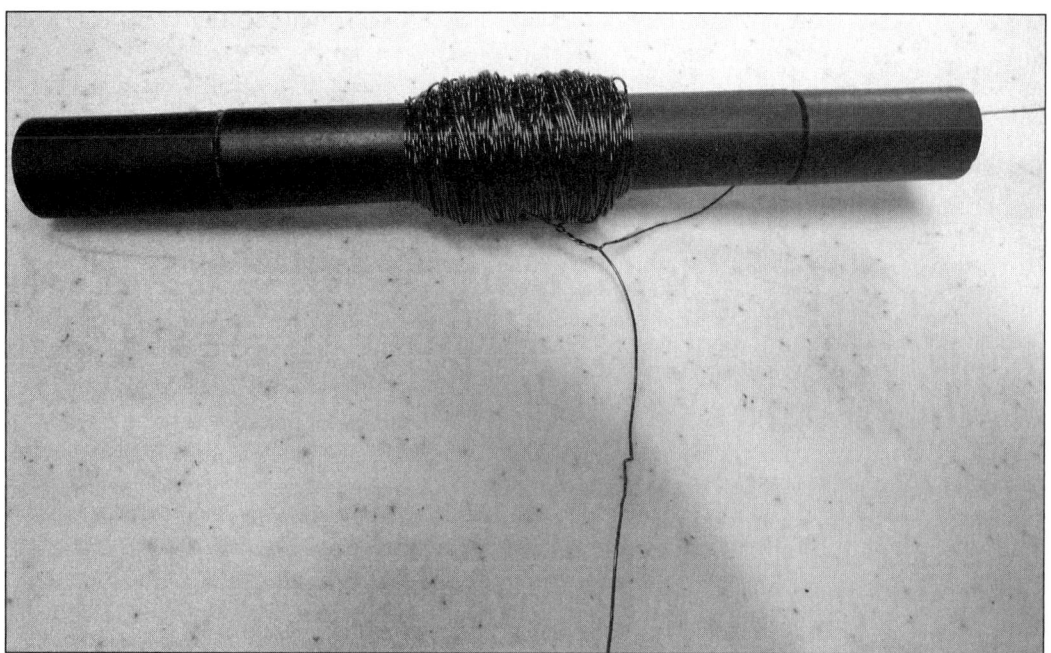

Figure 22.21 — Large ferrite loopstick made from several large ferrite beads. The antenna consists of 250 turns of no. 28 AWG magnet wire close-wound around the center of the rod. There's no need to wind the coil in neat layers.

antennas, I work with what I have on hand, and have always (well *almost* always) been gratified with the results. I have miles and miles of magnet wire on hand, and just a few yards of Litz wire. In my estimation, Litz wire is really just gilding the lily. Again, I welcome any verifiable reports to the contrary.

Since the 630 meter band is *so* narrow, there isn't much of an incentive to making the thing tunable, though the use of a small capacitor across the winding can easily give you a traditional tunable loopstick. I found it to be a simple matter to just sequentially prune the length of the wire until I found it to be self-resonant in the middle of the 630 meter band, just as I did with the frame loop antenna described earlier that I pruned to be self-resonant in the 2200 meter band. Having spent decades winding and unwinding coils, I don't find the process tedious at all. Surely many of my fellow hams have differing opinions on this matter. For those timid souls, there are countless formularies for figuring out the precise number of turns necessary for a given diameter, length and Zodiac sign of the ferrite rod in question. After applying the formulas, the thing will undoubtedly need tweaking for final tuning anyway. I prefer to use my accumulated decades of experience and intuition in such matters right from the outset.

Once the loopstick is wound and pruned to resonance, the remaining task is to connect the amplifier to the coil. The preferred amplifier for this project is the two-AD8067 configuration in push-pull, as used in the eXOgon channel amplifiers (again, see Chapter 15 for more details). This will allow the loop to truly act as if nothing is connected to it, allowing it to function as if it were almost in free space. The output of the push-pull transformer can be unbalanced, with the low side firmly attached to the U-channel shield, as should be the shield of the transmission line (presumably coax).

You can mount the antenna on a small pedestal and set it upon your favorite receiver, where it can be rotated for best reception or best null. It might be beneficial, as well, to provide an elevation adjusting mount. Sometimes being able to tilt a ferrite loopstick will allow you to achieve a deeper null.

As with any of the antennas in this book, the ferrite loopstick antenna may readily be incorporated in to an array of such antennas for additional gain and directivity.

Project 8: A Newfangled Adcock Antenna

In its simplest form, the Adcock antenna is nothing more than two vertical antennas driven out of phase. It has a perfect null to signals arriving perpendicular to the plane of the antennas. In fact, nearly all modern tracking radars use some form of Adcock antenna, or vast arrays of Adcock antennas, to precisely locate the direction of a radio source, which is actually a reflected radar pulse. In its most common implementation, two orthogonal Adcock antennas are erected, to allow direction finding in all azimuthal directions.

The primary advantage of the Adcock antenna over many other types of direction finding loops is that it is largely "deaf" to signals arriving from high angles, which can be very disruptive to effective direction finding. The Adcock is particularly suited to low band direction finding, provided that you have enough room for four widely spaced vertical antennas. The *length* of the antenna elements can be reduced considerably using active antenna methods. Four widely spaced active whips certainly occupy a lot less "sky" than four widely spaced full sized verticals.

Unfortunately the azimuthal resolution of an Adcock antenna depends largely on the *spacing* between the elements. *Mutual coupling* between elements of a phased array can make it difficult to obtain an ideal perfect null...though in theory it should be *possible* to obtain a perfect null with any two out-of-phase verticals. It boils down to the issue of how good is good enough.

But here's the great news. As it turns out, the shorter two nearby antennas are, the *less* the degree of mutual coupling! Of course, this shouldn't be too surprising. What this means is, for all practical purposes, we can decrease the spacing (almost) proportionally to the degree that we shrink each element, for a given degree of accuracy.

But now for some more bad news. In order for an Adcock antenna to perform properly, the phase angle of the two elements must be very closely maintained. This is not much of a challenge for passive arrays, where the phasing can be achieved with a simple coaxial T feeding two equal length transmission lines. (The 180° phase shift can be achieved by simply swapping the polarity of one of the antenna connections.) With an active antenna, the phasing is not so simple, as the active devices have a built-in phase shift, which may be largely unknown, or perhaps even unstable. What to do, what to do?

The simple answer is rather than wiring the antennas out of phase, feed them *in phase* into the differential inputs of a single op-amp, such as (what else?) the AD8067, which will then produce an output signal proportional to the phase shift of the antenna elements. Any incidental phase shift created by the op-amp or associated circuitry will be applied to both antennas, cancelling out all error (once original balance has been set up). Two such arrays, assembled orthogonally will give you nulls in four directions.

Now, as in any symmetrical nulling antenna, there is going to be ambiguity. An Adcock antenna will give you the same "answer" to signals 180° off from where you think they are, as well as the real thing. So you will need to provide some kind of sense antenna. For a "four square" Adcock array, the logical place for the sense antenna is right in the center. (See sidebar, "Making Sense of Sense Antennas.")

Making Sense of Sense Antennas

If you've followed this tome clear to Chapter 22, you have probably sensed that radio direction finding is a significant application of receiving antennas. And you would be correct in inferring this.

When it comes to precise direction finding, the *null* in an antenna pattern is significantly more accurate than the main lobe of, say, a high gain Yagi antenna. Nearly all effective DF antennas rely on nulling methods. All modern tracking radars use nulls to precisely locate targets, while major lobes are generally designed to *find* the targets in the first place.

The primary disadvantage of most nulling methods is that they are ambiguous; they usually achieve two nulls 180° apart. This, of course, is not normally the problem with a "standard" directional antenna which usually has a well-defined "front" and "back."

One widely accepted method for eliminating the ambiguity of a nulling antenna is to *temporarily* alter the bidirectional pattern into a unidirectional one. Typically, a *sense antenna* is an omnidirectional antenna, the output of which is combined with the bidirectional nulling pattern to achieve a *cardioid* pattern. In operation, the sense antenna is switched in to achieve a general idea of the direction of the signal in question (within 180°), which allows the operator to identify and eliminate the wrong heading.

Now, the question often arises as to why you can't just use a cardioid pattern to begin with. When properly adjusted, a cardioid is capable of an infinitely deep "notch" in one direction, just as narrow as that of a bidirectional null of a small loop or Adcock antenna.

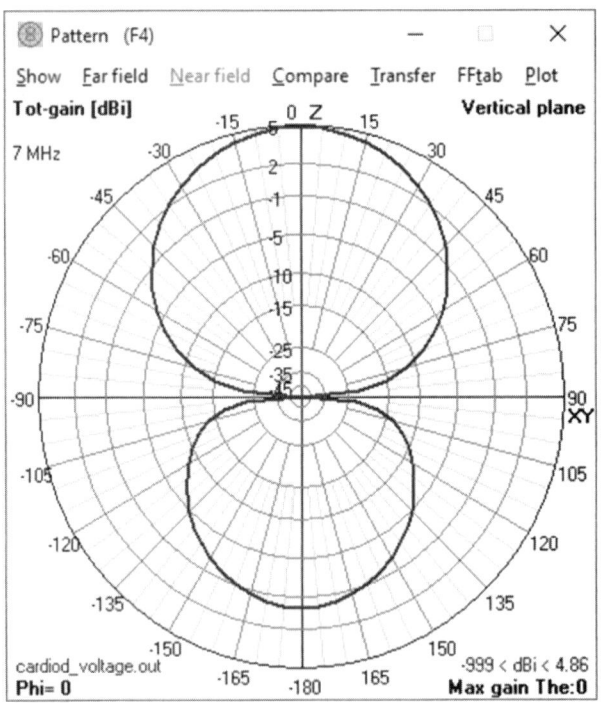

Figure 22.A — Dipoles with ¼ wave spacing and 90° *voltage* phasing

The operative phrase here is "properly adjusted." As it turns out it is infinitely more difficult to obtain a perfect cardioid null than it is to get a perfect null with a bidirectional phased array.

The ideal method for generating a perfect cardioid pattern is to space two dipoles ¼ wave apart and feed the *currents* precisely 90 electrical degrees apart. The first part is simple; the second part is much easier said than done. When a pair of dipoles are fed precisely 180° out of phase in *voltage*, as long as the array is physically symmetrical, the *currents* will also be precisely 180° out of phase. Any mutual coupling is entirely reciprocal, and all error cancels out. This is not the case with a cardioid array. It is a simple matter to design a phasing harness with a specific voltage phase, but very difficult to design a similar phasing harness to achieve a specific current phasing.

Notice the glaring difference between the two "cardioid" patterns in **Figure 22.A** and **Figure 22.B**. With a little minor tweaking of phasing or spacing, the second pattern can be nudged into a perfect cardioid, but as you can see the first example doesn't even resemble a cardioid. Even in the best case scenario, the pattern of the cardioid is *extremely* touchy; a minor change of *any* parameter will make the pattern unrecognizable. This is not the case with a bidirectional antenna, which is eminently forgiving.

So, in order to avoid endless frustration trying to achieve a perfect cardioid, the best practice is to just use a sense antenna to "warp" the pattern of a bidirectional antenna just enough to identify the "front" end of the thing.

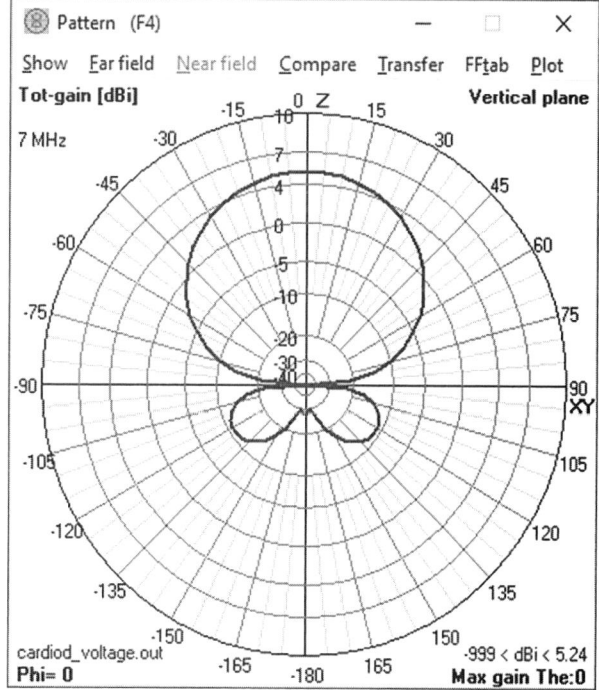

Figure 22.B — Dipoles with ¼ wave spacing and 90° *current* phasing

Project 9: An S-Meter for Direct Conversion Receivers

The final two projects in this book are actually closely related. At the core of these novel projects is the Binaural I-Q Receiver, with some modifications, described in March 1999 *QST*, by Rick Campbell, KK7B, and in many editions of the *ARRL Handbook*. The article is also available to ARRL members from the *QST* article archives on the ARRL website.

The excellent direct conversion (DC) receiver described in that article is actually a very slight variation of one of the most important instruments in all modern radio science, the lock-in amplifier. The principle (and block diagram) of the I-Q receiver and the lock-in amplifier are identical — only the actual values are different. **Figure 22.22** shows the block diagram of KK7B's I-Q binaural receiver and **Figure 22.23** shows the schematic, both reproduced from the original article.

One of the problems of the direct conversion receiver, the mainstay of scientific radio, is the lack of a convenient AGC and S-meter. In reality, it is the *lack* of AGC which makes the lock-in amplifier such a reliable instrument: what you see is what you get. No radio receiver is as

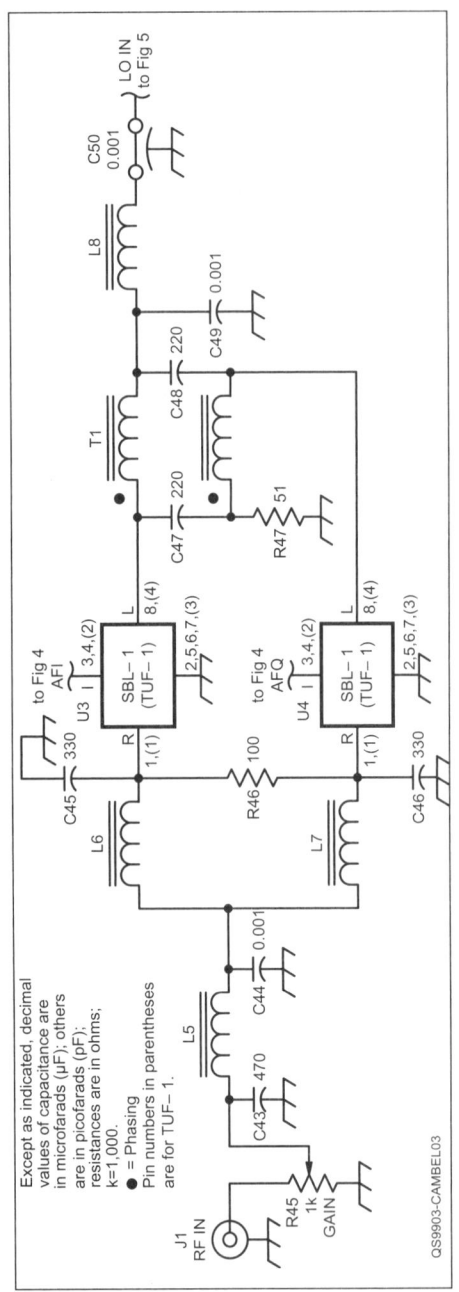

Figure 22.23 — Schematic diagram of KK7B's I-Q binaural receiver.

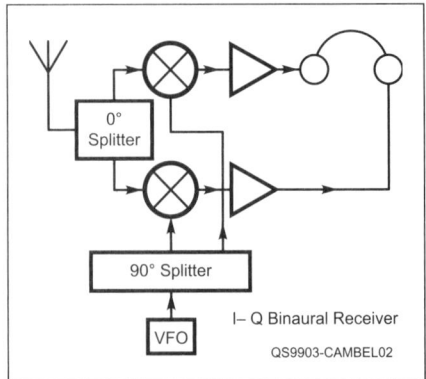

Figure 22.22 — Block diagram of KK7B's I-Q binaural receiver.

transparent as the direct conversion receiver to incoming radio signals.

In a single channel DC receiver, the output from the mixer is both amplitude and phase dependent. Although there is a dc voltage component from the mixer, that dc voltage is somewhat meaningless unless the local oscillator (LO) is phase-locked to the off-the-air signal. In the I-Q receiver, however, the *combined* (algebraically summed) dc voltage from both the I (in-phase) and Q (quadrature) mixers is proportional only to the *amplitude* of the incoming signal. In a lab grade lock-in amplifier, this combined dc voltage is normally termed the *magnitude channel*. The magnitude channel is always proportional to the incoming signal strength…as well as being extremely linear.

Figure 22.23 shows the quadrature detectors. All that is necessary is to tap off from the points labeled AFI and AFQ, and combine them in a summing amplifier. There's nothing critical or profound about this summing amplifier. A garden variety 741 circuit is more than adequate (**Figure 22.24**).

The output of this amplifier is inverted, so it should be followed by a simple inverting amplifier. The output of the inverting amplifier can be directly connected to a milliammeter of convenient scale. Again, this output will be very linear and directly proportional to input signal voltage. If it is desired to use a conventional S-meter, the inverter can be followed with a logarithmic amplifier to simulate the AGC action of a "normal" receiver. Refer to the field strength meter project at the beginning of this chapter for log-amp hints.

I cannot overstate the amazing capabilities of the I-Q receiver. We trust this simple project will add just a touch to your operating enjoyment of this fine instrument. But there's an ulterior motive here, as this is merely the stepping stone to a much more ambitious project, centered around Mr. Campbell's excellent work.

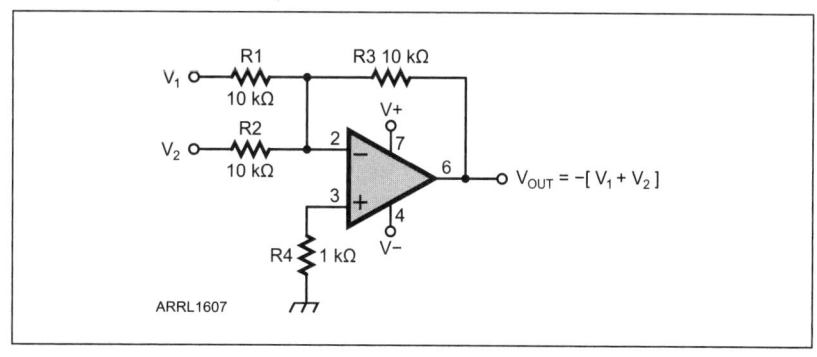

Figure 22.24 — Summing amplifier using a 741 op-amp. The output (V_{out}) can be connected to an inverting amplifier and milliammeter to monitor signal strength.

Project 10: A CPOL Diversity Array Using Three Antennas

And now we come to the *pièce derésistance* of this project chapter. No, that's not Latin for "piece of resistance," but rather a magnum opus of sorts for the ambitious receiving antenna experimenter. Are you ready to build *three* of Rick Campbell's Binaural I-Q receivers? This project consists of a triangular array of circularly polarized antennas, providing spatial diversity in two dimensions. This project serves as a powerful scientific instrument, as well as a practical enhancement to HF reception.

Now, don't be too daunted by the prospect of having to build three of these excellent receivers. The circuit will be greatly simplified because we can use a *single* local oscillator for all three receivers. Further simplifying matters, we will use a *crystal controlled* local oscillator, specifically 10 MHz, the design of which is vastly simpler than trying to build a stable VFO. (The crystal oscillator can be replaced with a stable function or signal generator, if you happen to have one in the shack.) Why 10 MHz? Because 10 MHz WWV is an ideal beacon station for observing X and O modes...and the system can be used for 30 meter "normal" ham operation with very slight modification.

Which brings us to one other refinement of the original I-Q receiver project. Instead of using three expensive 90° hybrids to obtain circular polarization, we will use the I-Q receivers to do the work. In the original circuit, antenna inputs are split into two signals, one feeding the I channel, and one feeding the Q channel. We will bypass the splitters, and feed the N-S channel(s) directly into the I mixer input(s), and the E-W channel(s) directly into the Q mixer input(s). CPOL sense switching can be easily achieved by a simply swapping the I and Q input channels. Again, this can be done manually, or a bit more elegantly with some relays or PIN diode switches.

The antenna array itself consists of three crossed (orthogonal) inverted V antennas spaced as far apart as property or economics permit. The antennas should be situated in an equilateral triangle, but the actual compass orientation is unimportant. (This is the beauty of using three rather than four antennas). A single 15-foot mast of 4-inch-diameter PVC plumbing is used for each of the crossed inverted Vs. It is recommended, but not absolutely necessary, that each inverted V has a balun at the feed point. All six transmission lines should be of equal length.

Each of the three I-Q receivers should be fitted with an S-meter as described in the previous project. This is important to show just how effective the diversity system is working, as well as to show the profound differences between X and O propagation.

Finally, all three I channel audio outputs (after bandpass filtering) are combined into a summing amplifier. The same is done with all three Q channel audio outputs — see **Figure 22.25**. **Figure 22.26** shows the block diagram for the complete system.

Figure 22.25 — Combining the I and Q audio channels from three Binaural I-Q receivers.

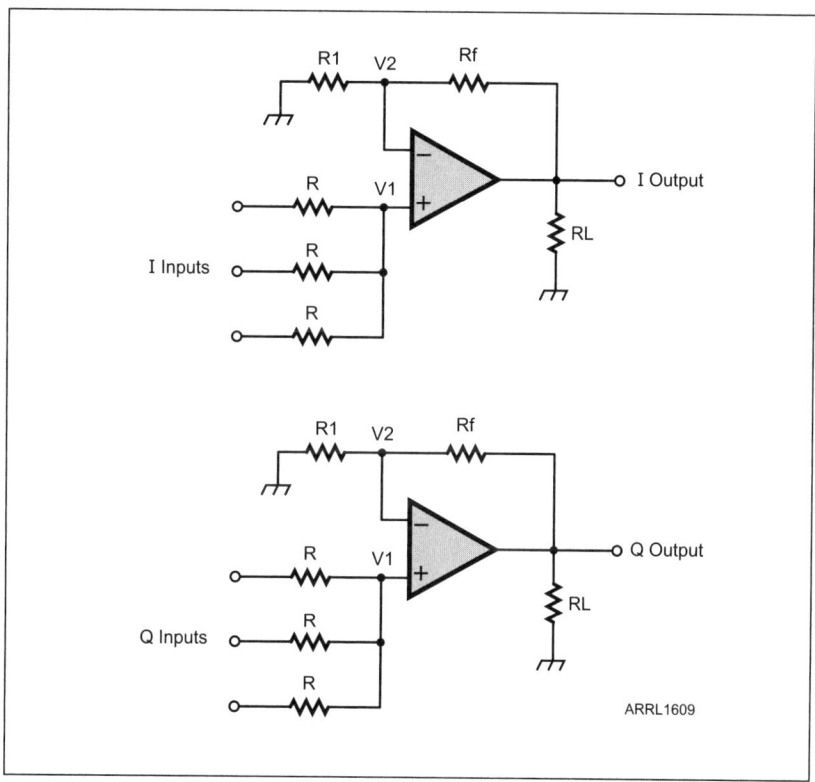

Figure 22.26 — Block diagram of the complete array.

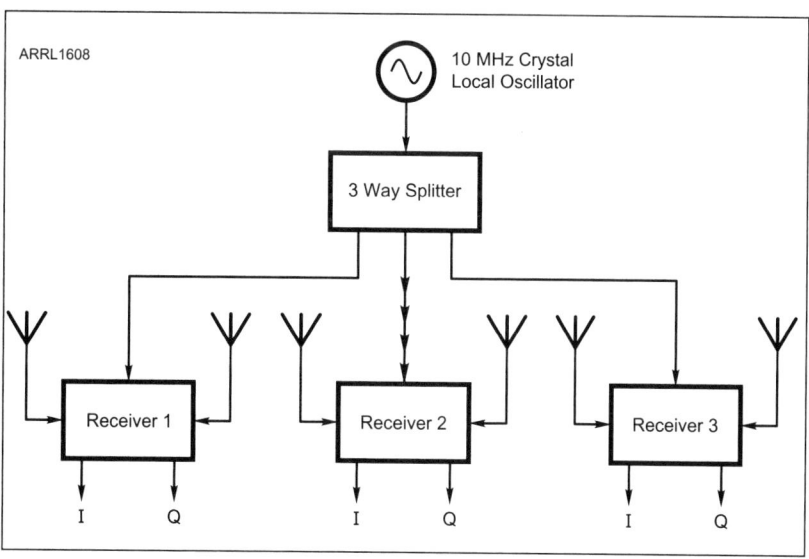

Well, that's about it. Like all the projects described in this book, this is only representative and can be readily "seasoned to taste" to accommodate any band or bands you might desire. This is a versatile instrument and can reveal many things about HF propagation that you are unlikely to have encountered before.

Closing Remarks

Any or all of the projects in this chapter can be combined in countless combinations to suit your needs or interests. Again, we have kept the details flexible and generic, in order to demonstrate some universal principles, as well as give you a wide variety of receiving antenna possibilities. More specific details can be addressed individually, and we welcome and encourage feedback and suggestions for improvements. The important thing is that you get on the air and experiment.

Chapter 23

Materials and Construction Techniques

One of the nice things about receiving antennas is that they don't have to be anywhere near as physically robust as transmitting antennas. This is doubly convenient when working with *experimental* antennas which may take a number of forms during design and evaluation before the final form is "chiseled in granite."

Power ratings of receiving antenna components are essentially nonexistent, in comparison with transmitting antennas. Power levels encountered are on the order of *picowatts* or less, compared to potentially kilowatts in transmitting applications. Effective receiving loops can be wound with wire not much thicker than a human hair (if you have any to compare it with), which would immediately *evaporate* with normal transmitting powers.

Of course, many receiving antennas, such as long wires and Beverage antennas, consist of long horizontal runs of wire, and such wire needs to be at least strong enough to not break under its own weight. In such installations, it's a good idea to consider the breaking strength of the wire. In addition, you might eventually decide that you want to use a successful receiving long wire antenna as a *transmitting* antenna too, in which case you want to be sure it doesn't evaporate the first time you hit the key. For small "concentrated" antennas such as frame loops, the mechanical strength of the wire is pretty much irrelevant.

The following is a list of materials you will want to have on hand for receiving antenna experimentation.

Magnet Wire...Lots of It

This is small gauge copper wire with very thin varnish insulation. Absolutely necessary for winding many-turn antennas and inductor components. It's nice to have a few gauges at your disposal. I have spools of no. 24, 48, and even some *64* AWG gauge wire...the last of which is extremely difficult to handle without breaking. It also tends to evaporate at the touch of a soldering iron...so the uses for this are *very* specialized. However, I do use a lot of no. 48 AWG magnet wire. Now, for most of us "properly aged" hams, magnet wire is one of those things you just *have*; it tends to spontaneously multiply, like wire coat hangers in the corner closet. I can't say I remember ever buying magnet wire; it's just an integral part of my half-century accumulation of radio parts. But if you're just entering the hobby, you may not even have even the smallest morsel of magnet wire capable of spawning more of the same in your parts drawer. So, you may find yourself having to actually buy some from some place such as **mwswire.com**. They have more kinds of magnet wire than you can shake a dipole at, and a ton of information about wire that you probably never thought of.

Buying magnet wire in bulk can seem a bit expensive, but I can almost guarantee that you'll never have enough of the stuff once you start building antennas (a rather addictive sub-hobby of ham radio). Get all you can afford…if nothing else, you'll have a stash you can retail to your fellow hams as they start building antennas too.

Project Boxes

Since many of the antennas in this book are *active* antennas, you'll want a secure and weatherproof place to mount your antennas' active components. I have a wide variety of aluminum "Bud Boxes," and a nice collection of plastic project boxes which are still (astonishingly) available at our local vestige of a RadioShack store. I'm also a good friend of the nice lady who owns the local tea house, and she supplies me almost countless various sizes and shapes of tea tins which are great for building radio accessories into (I showed one of these in Chapter 19). Other handy enclosures are pill bottles (which make great coil forms too). You can also build enclosures out of double-sided copper-clad PC boards, a recent favorite of mine. They just look classier and more expensive than aluminum chassis boxes, but take a little more skill in assembling, especially if you've never soldered anything with a significant amount of surface area.

Plastic household bins and containers are a perennial favorite for outdoor enclosures, especially for temporary and experimental situations (**Figure 23.1**). (At HIPAS Observatory we had dozens of Tupperware enclosed "temporary" scientific instruments out on the tundra that ended up remaining in the field for nearly a couple of decades. It wasn't pretty but it worked.)

"Re-purposed" instrument cabinets work great too. There's a lot of vintage (read "decrepit") equipment out there where the cabinets are far more valuable than any of the electronics currently residing within. I can't count the number of cabinets I've gutted for the purpose of making something more useful.

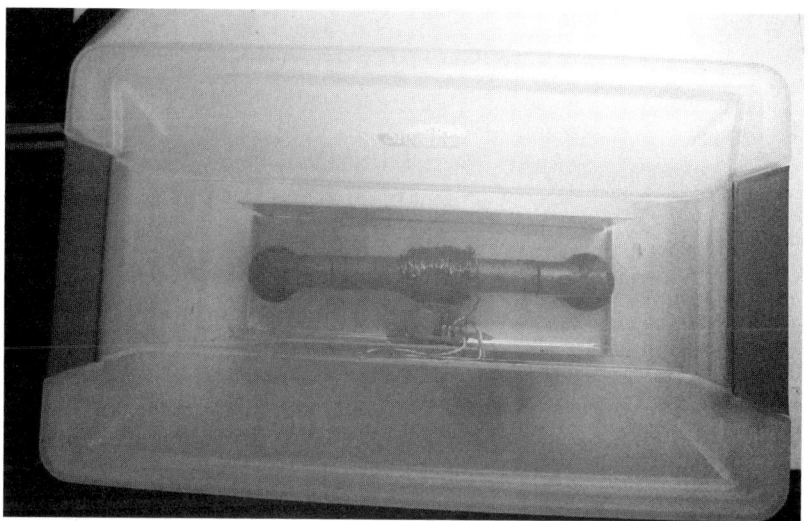

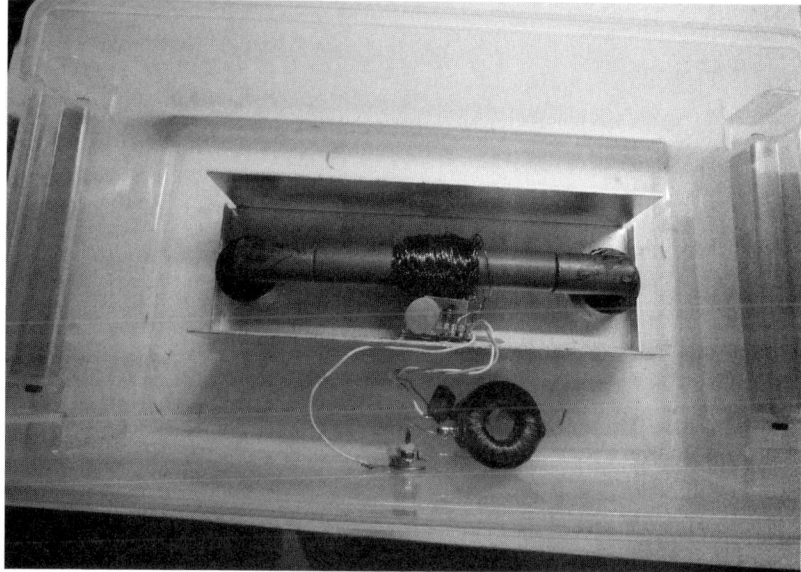

Figure 23.1 — Fresh from the kitchen. Plastic food storage containers make great weatherproof enclosures for antennas and other accessories. A shielded loopstick antenna is shown in both an open and sealed enclosure. The internal bias T is at the bottom, and a small PCB supporting the AD8067 preamp is mounted inside the shield. If necessary, the box can be lined with foam sheet insulation and heated with a small lamp if extra thermal stability is needed, such as when housing an oscillator of some kind. Here in Interior Alaska, we sometimes fill such enclosures with urethane spray foam, since we always have the glop lying around for sealing leaky windows and such. The finished product looks ugly, but it can last for decades. (Unlike a pristine, professional-looking unit, which is more likely than not to utterly fail the very first winter.)

Transmission Line

RG-59 or RG-6 75 Ω cable is most likely available in vast amounts as surplus from your friendly local cable TV outlet. It can't handle the power of RG-8 (or RG-58), but good cable TV coax is even better than those for reception, because the losses are lower. I can't think of any receiving antenna application where 75 Ω coax doesn't work just as well (or better) than the 50 Ω stuff. If you're really obsessive you can use transformers to match RG-59 or RG-6 to 50 Ω equipment, but there's seldom any point in this.

CAT5 cable (or any Ethernet cable, for that matter) is excellent for any receiving antenna transmission line. The eXOgon antenna, described in Chapter 15, uses CAT5 for both conveying the signals to the shack *and* powering the active amplifiers in the antenna. By the way, if you look closely at CAT5 cable, you'll see that each twisted pair has a slightly different "twist" to it. This is to reduce crosstalk within the cable when used for multiple signals. This also creates a very slight difference in propagation velocity for each pair — not enough to be significant, or even measureable, at HF, but an interesting point, nonetheless. At any rate, feel free to use inexpensive Ethernet cable with abandon in your receiving antenna projects. CAT5 twisted pairs have a nominal impedance of about 100 Ω.

Rope and Cord

I'm a great fan of "550 cord" sometimes known as "parachute cord" for both permanent and temporary wire antenna installations. Having worked as a military contractor for ages, I have a ready supply of the stuff, mainly from local DRMO (Defense Reutilization and Marketing Office) sales. (Alas, the DRMO auctions are not what they used to be.) I have had a number of 20-foot PVC masts using 550 cord as guy lines that have been up for *years*. Nylon rope can also be used, but it stretches terribly and deteriorates rapidly under ultraviolet.

Here's one use for cord you may not have thought of. I use 550 for long horizontal reaches as a "messenger" cable to suspend very light gauge copper wire. Thin copper wire stretches dramatically under its own weight, unless it happens to be *copper clad steel* ("Copperweld"). I despise Copperweld with a purple passion, though I have had to resort to using it on occasion. It's stiff, hard to work with, and if you live in a corrosive environment, such as near the beach where my first ham shack was, the copper cladding will be *gone* in short order. So, if you want to build a durable antenna with magnet wire, run a length of 550 cord first, and then use a series of cable ties every foot or so, to suspend the magnet wire from the cord.

Insulators

Small egg insulators seem to show up in vast quantities at hamfests and flea markets, and I usually pick up a handful when I encounter them. MFJ sells a new version of these for about 99 cents a pop. When properly installed, these can handle tremendous loads, as the forces are in *compression*. For lighter duty installations, any number of insulating materials are suitable. Remember that you don't even have to consider breakdown voltage in a receiving installation (unless you get hit by lightning, of course). As you've probably already gleaned, I use PVC for just about everything. A short sprig of PVC electrical conduit with a couple of holes drilled in it makes a perfectly suitable wire-end insulator.

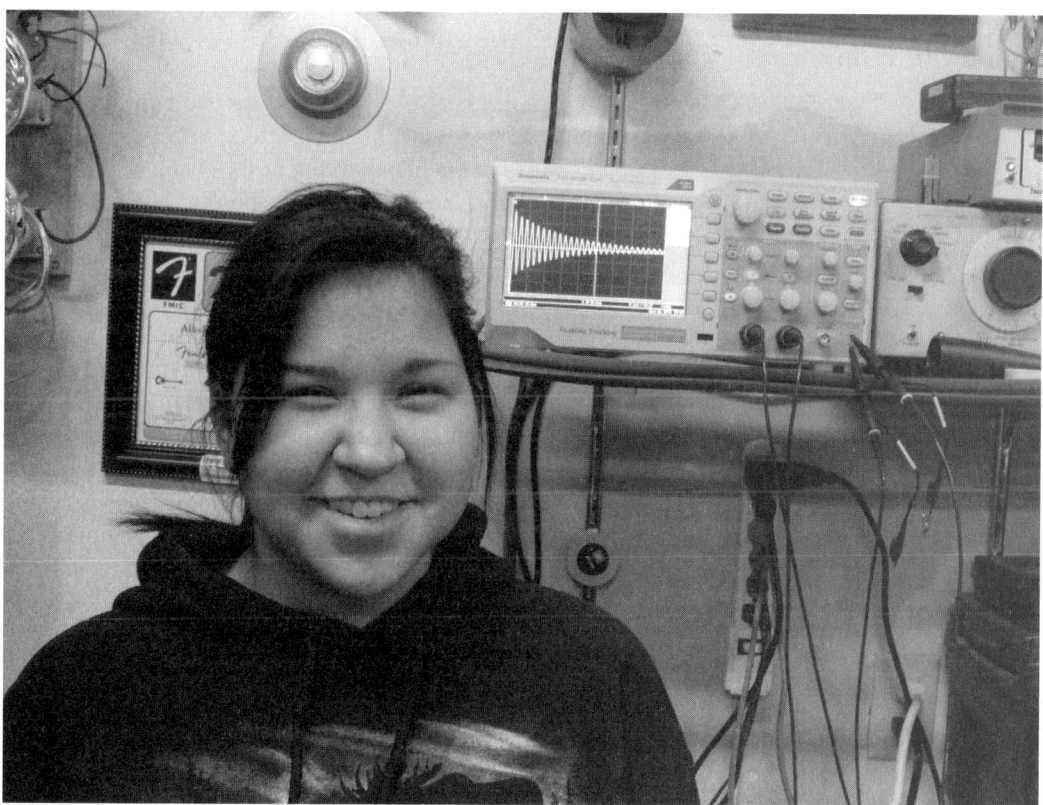

Figure 23.2 — Youth and homebrewing (now called "making") go together. Take every opportunity to teach a young person the "trade." Who knows if a few hours in the shop will launch them on a long fruitful career in electronics or radio. I had some great Elmers (mentors) in my formative years, and I trust that I've been able to inspire a few youths such as Jubilee Mountainflower to pursue science and technology.

Steel and Aluminum Rods

I have a large assortment of aluminum tubing, steel rods, and other long, pointy metallic objects, mainly for producing vertical whip antennas of various sizes.

Large PVC Plumbing

Two, three, and four-inch-diameter PVC pipes make excellent rapid-deployment masts for wire antennas such as K9AY loops. Smaller PVC tubes, such as electrical conduit, make great spreaders for frame loop antennas or cross-arms for turnstile antennas, such as the eXOgon. (One fellow I know actually erected a *40-foot* mast from 3-inch PVC plumbing, but I don't recommend it. I have no idea how he actually got the thing up there!)

Soldering Equipment

Of course, every ham needs a soldering iron…or two…or three. For general construction such as attaching connectors, a 35 W soldering pencil works fine. For working outside, nothing beats a butane soldering torch, especially if you're up in the air someplace. For fine work, you're going to want to bite the bullet and get an actual temperature regulated *soldering station*. Now, I've even done surface mount soldering with a "free range" soldering pencil a number of times, but this is not recommended. If you work with active antennas at all, you *will* be doing some surface mount soldering, so get the right equipment…and the right skills. By the way, I highly recommend getting a *binocular magnifier* for doing surface mount work. You'll be glad you did.

Now, if you still find the prospect of doing surface mount soldering too daunting, the next best solution (or perhaps the best solution) is to find some ambitious child labor to do the task (**Figure 23.2**). My young friend Jubilee "High Quality Eskimo" Mountainflower did all the surface mount soldering for my early eXOgon prototypes, and I occasionally still recruit her to do some fine soldering when I'm feeling too decrepit. She does this with *no* magnification whatsoever! (I don't think I would have *ever* been able to do this, even as a "yute" and I've always had good vision.) In any case, it's always a good thing to get young folks involved in ham radio in any way possible.

PCB (Printed Circuit Board) "Fixin's"

For one-off projects, it's probably not a bad idea to learn how to etch a circuit board. The *ARRL Handbook* has a good section on this not-so-lost art. I find it a lot of fun, actually…it's one of the few activities I still do that involves roiling, vile potions. (Actually circuit etching solution is not *that* vile or roiling, but *do* use precautions such as safety goggles and gloves when working with the stuff.) I have a nice stash of single and double clad PCB material, both for making circuit boards and handsome, coppery cabinets. You can also get pre-etched circuit boards from a number of places. These come in a number of generic trace patterns that work for a large number of standard circuits such as op amps and TTL chips. These boards are fine for through-hole construction, but aren't normally suitable for surface mount work. If you still find yourself averse to using roiling chemical potions, another great way to etch a PCB is with a Dremel or similar rotary tool. Just use a high speed fine grinding or engraving bit to "erase" the copper where you don't want it. Be sure to use a particle mask in addition to goggle when doing this… copper dust and glass epoxy dust are not things you want to breathe.

Your "Normal" Tool Set

Over the years, the list of tools I deem *essential* has gradually grown (well, not so gradually, actually). I'm an inveterate tool junkie. But, here are some things you *absolutely* will need.
- A good pair of flush-cutting diagonal cutters
- A good pair of needle-nose pliers
- Flat-head and Phillips-head screwdrivers
- Small socket wrench or nut driver set
- A miter box and back saw (used mainly for cutting PVC tubing in my case)
- Variable speed drill and bits
- A Dremel tool (I used to consider this a nice optional widget, but I now consider it absolutely essential).
- A hammer of some kind. (One would think this would be a no-brainer, but you wouldn't believe how many hams I've met who can't even produce a hammer. Come on, guys; *ham*mer even starts with *ham*!)

Some Test Methods

Figure 23.3 shows my workbench. Going back through the deep dark history of radio practice, it's amazing to find what hams (and SWLs) we were able to accomplish with virtually no test equipment. We have a vast array of excellent, very affordable test gear that hams of yore would have died to have. But the fact remains that having some good test and analytical skills is better than the best test equipment you can buy.

Figure 23.3 — The main R&D workbench in the garage of KL7AJ features an eclectic assemblage of trusty old and sparkling new test equipment, under the constant supervision of the Ohm Gnome (far right). You don't need a lot of fancy equipment to do a lot of fancy stuff with antennas. Always remember, Amateur Radio is something you learn, not something you buy!

Most of the *broadband* antennas described in this book are very forgiving in their setup and tuning. In fact, there *is* no tuning involved in the classic active antenna...that's one of the main points of the thing.

However, a number of the antennas described herein, such as small loops, have very *sharp* tuning, which, while not requiring any exotic test equipment, *does* require some good practices.

A dip meter (originally called a grid dip oscillator, or GDO, back when they used tubes) is wonderful for tuning up resonant loop antennas. For decades, if an impoverished ham had any test equipment at all, it would most likely have been a GDO...even ahead of having a voltmeter or ohmmeter. Unfortunately there aren't readily available dip meters that work down in our new lowest frequency bands, at least yet. (This might be a good product for an enterprising ham to produce.) And, this frequency range is even below the range of most amateur antenna analyzers.

For this kind of measurement, I use an oscilloscope and a function generator, both of which are available at incredibly reasonable prices. I currently have a rather newfangled digital storage oscilloscope (DSO). I wrote a whole book about these gems, by the way, and it's available as an eBook on Amazon. It's a great book if I might say so myself. But even if I didn't say so, it's still a great book.

But long before I had my DSO, I had a long procession of old-school oscilloscopes, ranging from abysmal and decrepit to absolutely wonderful. I've gotten rid of most of the decrepit ones and only keep a couple of the wonderful old ones. At any rate, there really is no excuse for any ham not to know how to use an oscilloscope.

In **Figure 23.4**, I show the method of measuring the resonant frequency of a loopstick antenna. The oscilloscope probe is connected across the coil as you see. The INPUT from the function generator might be surprising to you, however. The ground and hot lead are clipped together

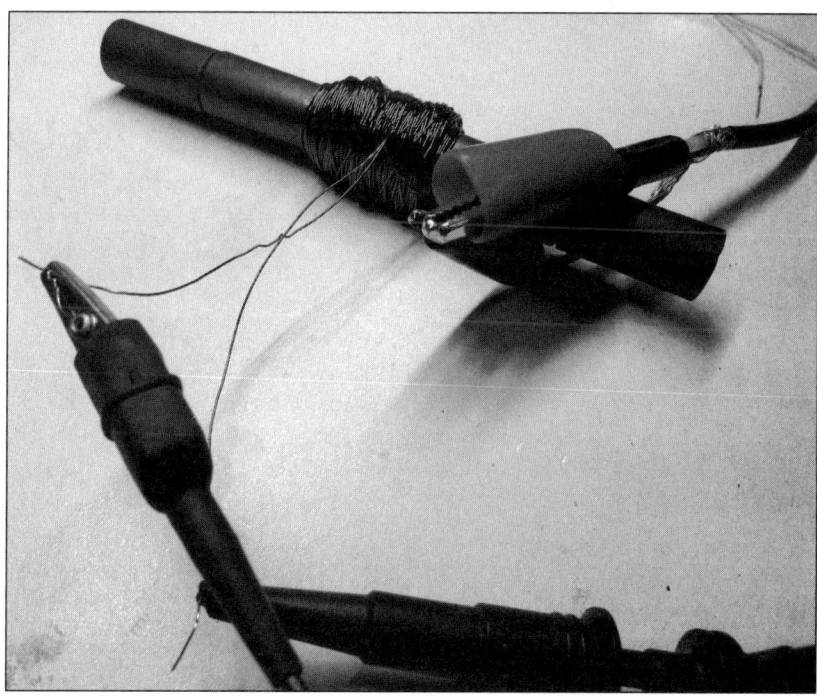

Figure 23.4 — **Measuring the resonant frequency of a loopstick antenna.**

Becoming Your Own RadioShack

One of the "excuses" a lot of hams come up with for not pursuing homebrewing with a vengeance is the lack of readily available radio parts. I am the first to admit that the parts market is nothing like it once was. I grew up in Silicon Valley, where there seemed to be an excellent electronics shop on every street corner. And every drugstore had a tube tester. And that once-Titanic entity known as RadioShack. Alas, RadioShack is no longer the Titan it once was, even in the places where it still exists. Its fate has pretty much followed that of the real Titanic.

So, the parts procurement landscape has definitely changed, but it hasn't gone away. One just has to be a little more resourceful. I found that I have had to become my very own RadioShack. And with a little dedication and ingenuity, you can too.

Now, I have to admit, I have had a good head start on the process…nearly a 50-year head start. I've been collecting electronics components ever since I knew what they were. I hardly ever have to buy anything. But if you're just starting out in the hobby, there are plenty of ways to start rapidly building up a great collection of parts bins. The first thing you need to fill your parts bins is…well…some parts bins! I've got a wall full of them as seen in **Figure 23.A**. Small parts bins can be obtained from home centers or online for a very reasonable cost. Our local hardware store has served me well in this regard. Your mileage may vary.

Electronics outfits such as DigiKey and Mouser sell resistor, capacitor, and inductor kits in pre-stocked bins. This is a great way to get a kick-start on your parts inventory. Also, check out eBay. Hundreds of electronics "grab bags" show up regularly. This is a wonderful source for things such as discrete transistors and diodes, as well as through-hole integrated circuits such as op-amps and TTL devices. Through-hole semiconductors are gradually being replaced by surface mount components, and some of the perennial favorites don't grow on trees like they used to. However there is still a *huge* new old stock (NOS)

Figure 23.A — Where have you bin all my life? One can never have too many parts bins…or parts to fill them with. Some persnickety hams actually go so far as to actually *label* said bins.

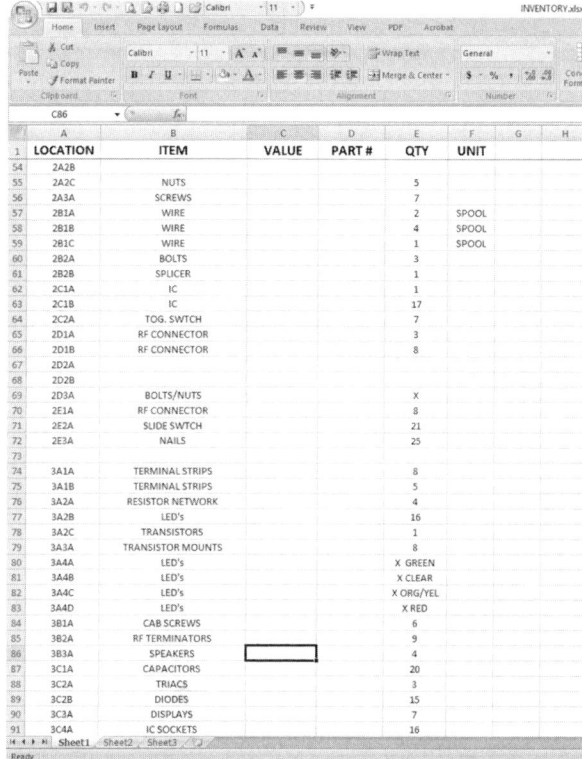

Figure 23.B — Or, you can go high tech and use a spreadsheet to keep track of all your parts. I am currently in the process of cataloguing and inventorying my collection of increasingly harder-to-get-parts accumulated over more than four decades.

inventory of such parts, so whenever you find them, grab them up in bulk, if possible. (I have accumulated about 10,000 germanium diodes through such sources, even though they have been considered by some to be "unobtainium." Of course, I may be part of the reason why they're unobtainium…but we won't go into that.)

The same goes for vacuum tubes. Of course, you probably won't be using many vacuum tubes for receiving purposes…but they do make for some great trading currency for components you *can* use. I am almost embarrassed to confess how many vacuum tubes I have, but it approaches five figures. I don't have them all in my shack…they're distributed in various friends' garages around the town. But I know exactly where they all are! Of course, this didn't happen overnight.

DRMO auctions are another great source of electronics components (**www.govliquidation.com**). However, you won't find as many complete radios at the auctions any more, as a lot of this equipment has or had classified encryption hardware in it. So you aren't likely to be able to pick up a surplus Harris PRC-160 manpack radio. (Alas.) But it is still absolutely astonishing how much NOS electronics is available via military surplus.

So, the bottom line here is that where there's a will, there's a way. Somebody somewhere has just the parts you need for that antenna project…and just about any other radio project you can think of. Get building!

forming a *single turn* loop. What this does is provide *very loose* coupling to the loopstick, so it maintains its Q, allowing an accurate resonance measurement.

If you don't have any experience with RF circuitry you're probably wondering how you could possibly get any coupling from the function generator to the main coil with this setup. In reality, this works great. This is normal and intuitive for experienced radio folks…the sort of thing you only get from lots of playing around with this stuff!

The 10:1 oscilloscope probe has 10 pF of capacitance, which forms a *small* fraction of the total circuit capacitance. You simply adjust the frequency of the function generator to obtain the maximum amplitude of the oscilloscope. You don't care what the actual amplitude is; all you're looking for is the frequency where this is maximum. (This is why you can get by with a really lousy oscilloscope.) If you *have* a frequency counter, you can get a little better frequency reading than the dial on your function generator has. Or, if you have a DSO, you can simply punch the FREQUENCY button and it will measure it for you. This is another advantage of the DSO over the geriatric oscilloscope.

I now use the function generator and oscilloscope for measuring all my tuned low band antennas. But I still haven't retired my trusty old grid dip oscillator. It is a reliable old friend for all my HF (and most of my VHF) antenna twiddling.

Radiation Resistance? What's That?

For most *transmitting* applications, the radiation resistance is equally as important as the resonant frequency of an antenna. In fact, it's actually *more* important, when it comes to efficiency.

You might have noticed, or even been chagrined by the fact that we seem to have ignored the resistive component in our antenna tuning discussions.

In a small receiving loop, or even a short active whip, the radiation resistance is essentially non-existent. It is always extremely low. With enough diligence, we *could* measure the radiation resistance of these antennas, but there would be little point. We don't need to know it because we aren't trying to *match* it to anything. A small, tuned loop operates with *far* less coupling than necessary for maximum power transfer. In fact, for best performance, we want to always approach the "loose coupler" model, which maintains the exceptionally high Q of the antenna. And we can get by with poor power transfer, because, at HF frequencies, we have far more gain than we can use. So we sacrifice efficient coupling to gain a high signal-to-noise ratio. And that is the final goal of any receiving antenna.

Chapter 24

Our Two New Bands

It has been said that timing is everything. While the veracity of that statement is probably debatable at times, it *is*, however, most gratifying to be finishing this book *just* as the final ribbons are being tied (legally speaking) on our two brand-new radio bands at 630 meters (472 – 479 kHz) and 2200 meters (135.7 – 137.8 kHz). These two new bands are *real* amateur bands now, fully under the auspices of FCC Part 97. There has also been a long-standing experimenter's band on 1750 meters, about 160 kHz. Denizens of this band have been referred to as "lowfers" and have accumulated a lot of experience and data on such low frequencies. They are hitting the ground running on 2200 meters.

While there is nothing *fundamentally* different about antennas on 630 and 2200 meters, as compared to those designed for our other lower frequency bands, there is definitely a difference in the *degree*. Practical antennas for these bands are *much* smaller, in terms of wavelength, and efficiency, than even 160 meters. To grossly oversimplify the matter, 630 meters is about four times as hard to do as 160, and 2200 meters is about 14 times as hard to do as 160.

Fortunately, *receiving* such low frequencies is infinitely simpler than *transmitting* on them. Well, I suppose *infinitely* more difficult might be overstating the case a bit...but not by much. The truth is that we have the easy part in this book; the really hard stuff has been covered by the ARRL 600 Meter Experimental Group (**500kc.com**), and the aforementioned "lowfers" who haven't received anywhere near the respect they deserve.

LF, VLF, and ELF have long, long history in terms of radio. In fact, some of the methods are so old that they've basically been *forgotten*...at least by most living radio amateurs. Our new bands afford us the wonderful opportunity to *relearn* some of these lost arts...and perhaps add a few new tricks in the process.

Just like AM Broadcasting, but Different

If you're an old geezer like me, you might have gotten your first taste of the magic of radio by means of a crystal radio. Although some hams did experiment with crystal radios on HF frequencies, the vast majority of crystal radio receivers were built to receive AM broadcast stations. If you're too young to have a clue what we're talking about, it's not your fault. Pay a visit to the Xtal Set Society (**www.midnightscience.net**) for some wonderful articles on crystal radios, historical and otherwise. In addition, you can get all kinds of vintage radio parts from these folks. I always have a great visit with them on my yearly pilgrimages to the Dayton Hamvention.

Our new band at 630 meters, more precisely 472 to 479 kHz, has propagation much like the lower end of the AM broadcast band. Now, what a lot of folks don't realize is that the AM broadcast band (525 to 1705 kHz) is *huge*. The top end of the broadcast band has *entirely* different propagation characteristics than the lower part of the band, being over three times the frequency. The top end of the band is very much like 160 meters, and is capable of impressive DX after nightfall. The lower end is much more restricted to ground wave propagation, though there are occasions when AM DXing can be had on the lower end of the band as well.

That being said, the bottom end of the AM band is a good indicator of how propagation will work on 630 meters. If you are hearing a lot of distant AM stations down around 500 kHz, it's likely you will be able to copy some 630 meter ham activity. But be aware that AM broadcast stations are likely to have tens of thousands of watts of ERP, whereas a 630 meter ham station will be doing well to achieve 5 W ERP. So, on the average, amateur 630 meter signals will be about 40 dB down from an equivalent "low-end" broadcast station.

Because such signals are so weak, special methods need to be applied to receive them. And, by extension, data rates will be *extremely* low. The traditional means of communication by very weak signals on the very low frequencies has been by very slow CW…known as QRSS. Tried and true methods of detecting these signals, such as the *lock-in amplifier* are very effective in decoding these still small voices buried far below the noise floor. There is a price to be paid in using the lock-in amplifier, however, and that is *time*. You need to be thinking in terms of symbols per *minute*, and bandwidths in *millihertz,* not symbols in kilobits per second and bandwidths of kilohertz!

Digital modes such as JT9 are also popular (**Figure 24.1**). These modes use modern digital signal processing techniques to decode signals well below the noise level. The software just requires a modern computer and sound card, and no special hardware is needed.

Highly selective, low-noise receiving antennas make the whole job

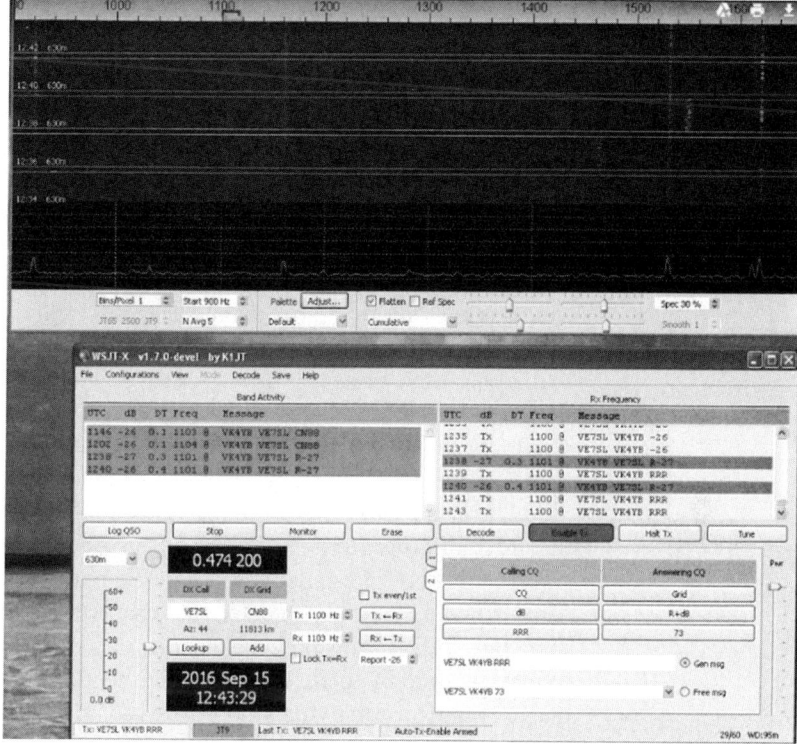

Figure 24.1 —A screen shot from a 630-meter JT9 contact between VE7SL in British Columbia and VK4YB in Australia.

a lot easier…or at least possible. Fortunately, interference from other stations is much less of a problem on our two new bands than on our other bands, at least for the time being. It is almost a given that relatively few intrepid souls will venture into the nether regions of the Amateur Radio spectrum…but we could be very wrong, and that could be a very good thing!

The Long and Short of It

If you have the real estate for it, a Beverage antenna might be the most effective antenna for 630 meters, while for 2200 meters, it is *highly* unlikely that you can build a Beverage big enough to come close to matching the performance of other available receiving antennas. As a rule of left big toe, you will need at least a wavelength for a Beverage to be very effective.

So unless you live on a 500,000 acre cattle ranch, or the Alaskan tundra, most likely you're going to be using a tuned loop of some sort. Even "small" loop antennas should be made as large as possible; you'll need all the gain you can possibly muster. And, as we've established, the gain of a receiving loop is proportional to its enclosed area. We don't suggest trying to "miniaturize" any receiving antenna for 630 or 2200 meters, in any way.

Hot Rods

Our new 2200 meter band officially falls into the "long wave" category. While different jurisdictions have different lines of demarcation between LF, VLF, ELF, UHF, and so on, we can safely include 2200 meters in the long wave class just about wherever we live. A wonderful resource for long wave reception techniques may be found at **www.vlf.it**. It is well worth your time to peruse the entire site; and there's plenty of weird science in there too!

Taking a cue from the "lowfer" lads, it *almost* goes without saying that the multi-turn ferrite rod is the way to go. Large ferrite loopsticks can be quite expensive, but you can glue a bunch of high permeability semi-large ferrite beads together end-to-end and come up with a really good facsimile. I show how this is done in Chapter 22, if you happened to have breezed by there. Be prepared to wind *hundreds* of turns of small gauge magnet wire around your ferrite rod! It's almost impossible to put too many turns on one of these — see **Figure 24.2**. (Commercial and scientific magnetometers typically have 100,000 or more turns on them,

Figure 24.2 — A ferrite loopstick antenna with many turns of small-gauge wire.

Figure 24.3 — Here a typical method of providing a low impedance output from a high impedance resonant loopstick is shown. The twisted ends of the high impedance winding form a "gimmick capacitor" which can be used for fine tuning this self-resonant loop. The number of turns for the low impedance winding is determined experimentally, usually the fewer the better, as this assures the least amount of loading of the resonant loop.

but they also operated down in the millihertz range. You don't have to be *quite* so radical, but you'll be in the same ballpark.)

Now, a ferrite rod wound in such a manner has *extremely* high output impedance. But that's not a problem now that we have devices such as the AD8067 FET op-amp at our disposal. Before such things existed, the standard practice was to wind a low impedance coupling loop of just a few turns around the main loopstick winding as shown in **Figure 24.3**. However, trying to figure out a "reasonable" number of turns for the secondary winding was (and is) always a tedious trial and error process. A lot of the trial and error can be eliminated by just assuming the primary, high-impedance coil has infinite output impedance, which matches nicely with our op-amp's (nearly) infinite input impedance. It makes things so much simpler.

However, one thing the op-amp *won't* fix for us is the fact that such an antenna can have *many* resonant frequencies, caused by the parasitic capacitance of all those close-spaced windings. Most of the time, a lot of extra resonances won't cause too much trouble...unless one of them happens to operate as a wave trap at precisely the frequency we want to receive! This can be resolved simply enough by moving the unwanted resonance around a bit by adding or subtracting a couple dozen turns or so. An antenna analyzer or vector network analyzer (VNA) can go a long way in identifying peculiar resonances in a ferrite loopstick...but doesn't do much to predict or avoid them before the fact. (By the way, excellent VNAs have become extremely affordable; there's little excuse for the dedicated ham experimenter not to have one.)

The bottom line is that any practical loopstick antenna *will* have parasitic resonances. You'll just have to deal with them. The benefits of that big chunk of ferrite far outweigh most disadvantages of *not* having one.

Speaking of Which

It is *not* always necessary to have a loopstick antenna tuned to resonance (usually by means of an external parallel capacitor). We mentioned the loopstick magnetometers which are always operated in a broadband, untuned mode. While resonating a loopstick increases the voltage at the terminals, the increased circulating current resulting from this can also lower the Q by ohmic heating (yes, this can occur at receiving power levels!), which actually can do exactly the opposite of what you want! This becomes a noticeable phenomenon when using extremely fine wire, where the dc resistance is already significant. Determining the precise parameters you need for optimum Q can be a daunting exercise, to say the least, and, at least in Amateur Radio circles, there's naturally very little data on the matter. This is an area ripe for a whole lot of experimentation.

Vertically Challenged

While we have strongly suggested that any practical communications on our new bands will be vertically polarized, since all ground wave signals *must* be vertically polarized, there might be one useful exception to this. NVIS signals (signals arriving from nearly directly overhead) will *not* need to be vertically polarized. In fact, such signals cannot, obviously, be even defined in terms of vertical or horizontal. So another avenue for exploration might be the reception of NVIS signals on our new bands.

Now, already some of you are protesting that D-layer absorption is going to totally eliminate any long wave NVIS from happening in the first place. Under most normal circumstances, this would be true. But it wouldn't necessarily apply to signals that *originated* above the D layer, late at night…such as from a possible satellite or Space Station experiment. Or possibly from some *chordal hop* path. There are still a lot of unknown possibilities here. And again, our new bands will be a wonderful playground for exploring these!

Our new bands were hard won! I strongly urge everyone to at least dip your toe in the pond. As the higher frequencies deteriorate throughout the rest of the current sunspot cycle, we may be very glad we have these new bands. Enjoy!

Appendix A
The Q of Everything

It really is astonishing how consistent the physical universe is. We can readily exchange countless principles among electronic, physical, mechanical, chemical, and sometimes even biological disciplines. This is a good thing, because it allows us to use clear, obvious, and familiar mechanical analogies to explain things in the electronics world that are not so clear, obvious, or familiar.

One great example of this is the concept of Q, of which we've already spoken at length. Although the topic is not too complicated, at least from a mathematical standpoint, I've found that even many active hams have a fuzzy and foggy concept of Q when it comes to electronic circuits. So, it's helpful to use some "folksy" analogies when explaining the topic.

In any physical system, we can do two things with energy. We can either dissipate it in the form of heat (sometimes with an intermediate process or two involved), or we can store it in some container for later use. One of the most familiar physical systems is a simple pendulum, a weight swinging at the end of a string. If we were to create a pendulum in a perfect vacuum, (along with a perfectly flexible string), and gave it a nudge, the pendulum would swing back and forth forever at a precise frequency determined only by the length of the string.

In the process of swinging back and forth, the pendulum is "sloshing" energy back and forth between two "containers." One of these containers is potential energy, which is a function of the height of the pendulum. The other container is kinetic energy, which is a function of the velocity of the pendulum. At the top of the swing, the kinetic energy is zero, while the potential energy is maximum; at the bottom of the swing, the potential energy is zero, while the kinetic energy is maximum. The *change* of potential energy and kinetic energy follow a sinusoidal pattern, not too surprisingly. (Technically, this pure sinusoidal function only exists for low levels of swing, but it's certainly close enough for government work.)

One remaining factor of the swinging pendulum is the *mass*, which really only becomes relevant when there is some friction involved. As long as there is no friction in the system, the pendulum will swing indefinitely at its *resonant frequency*, again determined *only* by the length of the pendulum. Such a pendulum has a Q of infinity.

Now, if we have any friction in the system, let's say our bell jar containing the pendulum has a small leak), the pendulum will lose some energy with every swing. This lost energy is converted into *heat*, raising the temperature of the air molecules by a minuscule amount. Now, with any friction at all, we know that our pendulum will continually lose energy. However, if we increase the *mass* of the pendulum (by using a heavier weight), the loss of energy will be more gradual. In reality, we've actually started with more energy with a heavier weight, for a given initial height of swing.

So, how do we translate this into electrical circuitry? Well, the most obvious parallel is that friction is equivalent to electrical resistance. We know that resistors get hot and they dissipate heat. They burn up electrical energy.

A slightly trickier translation would be *mass*. What is the electrical equivalent of mass? Well, we *could* take an ultra-literal approach and consider the total mass of electrons we're sloshing around in a resonant electrical circuit. If we slosh a lot of electrons back and forth, we have more energy to work with. This is not a perfect analogy, but it's a start.

Now, let's talk about the storage containers. In the case of the pendulum, one container is kinetic energy, and the other container is potential energy. In a resonant electrical circuit, our two containers are capacitance and inductance. (To put a finer edge on it, the containers are actually the electrical and magnetic *fields,* but capacitors and inductors have better "handles" to work with. Our resonant circuit sloshes energy back and forth between these two containers, and if there is no friction (resistance), the sloshing around will continue forever. Such a circuit would have infinite Q.

Now, when friction or resistance is added to a simple resonant circuit, not only do we lose energy with every "swing" or "slosh,", but the actual resonant frequency becomes less narrowly- defined. Let's take an extreme case to illustrate this. Imagine a very low Q pendulum: a one-foot long string with a knot tied at the end, serving as the pendulum bob, operating at atmospheric pressure. It would be nearly impossible to define the frequency of resonance; all the energy is pretty well gone by the first swing. This would be a circuit of essentially zero Q. In this case the energy dissipation far exceeds any energy storage.

This same analogy can also be applied to other systems, such as satellites in orbit. An orbiting moon can be thought of as a resonant circuit of high Q; while a low-orbiting satellite gradually burning up in the atmosphere would be a rather low-Q circuit.

There's a lot more that could be said about this topic. We can also describe a resonant circuit in terms of its "inverse-Q" or damping factor. You might want to check out my ARRL e-book, *Digital Storage Oscilloscopes for Ham Radio* for a detailed description and some beautiful displays of damped oscillations.

I hope this little detour has helped someone more fully understand this important and fascinating concept of Q.

Appendix B

Get a Load of This: Taking the Mystery Out of Loaded and Unloaded Q

We are usually (and rightfully) skeptical of any article that begins, "Taking the mystery out of..." because it is almost inevitably a harbinger of even greater confusion. However, we can (almost) assure you that after carefully considering this article, you will have a comfortable understanding of the matter.

The issue of loaded and unloaded Q has only an indirect connection with antennas, being far more applicable to things like transmitter Pi-networks and antenna tuners, but it's a rich enough topic to include here. You may find it has more to do with antennas than you may have thought.

To fully clear the haze surrounding the subject *does* require that you understand — and more importantly — absolutely *believe* Kirchhoff's Current Law (KCL).

Simply stated, KCL tells us that the sum of all the currents entering a *node* has to equal the sum of the currents *leaving* the node. If we assign a *direction* or *sign* to each of the currents, the law can be even more simply stated as "the sum of all currents passing through a node is zero."

This law, on the surface, seems to state the obvious. In terms of plumbing, it says that you have to have the same amount of water leaving one end of a hose as you put into the other end. (Duh!)

Yet, despite this incredible simplicity of KCL, the implications are profound, and, depending on your mindset, can lead to either hopeless confusion (Say what?!) or extreme enlightenment (Aha!). We trust you prefer Aha! to Say what?!

You Node This All Along

First, let's define a node as connection between any two (or more) electrical components. If you work with *SPICE* circuit simulation software, the concept of nodes should be familiar. The *simplest* electric circuit, a voltage source and a resistance, has two nodes.

Here's another semi-law sort of statement: Any complete circuit must have at least as many nodes as the number of components. This should be another no-brainer.

Now, let's take our simplest *RLC circuit*, as shown below. (Question marks, such as R1 ???, simply indicate the actual value is unspecified).

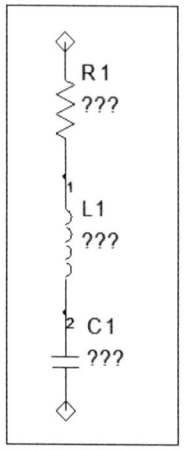

As you see, it has two nodes, labeled 1 and 2 (Node 1 is between R1 and L1, and Node 2 is between L1 and C1). But there are three components. Guess what? It's not a circuit!

Now, at first blush, probably 90% of radio amateurs would look at this and say it's a *series* resonant circuit. But it's not. It's not a circuit at all. So how do we make it a circuit? We have to connect the top and bottom of it together. Let's go ahead and do that.

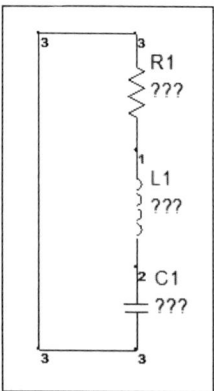

Well, now we have three components and three nodes. It's a complete circuit. But is it a series circuit, or a parallel circuit? Just to make things a little simpler, let's temporarily get rid of R1. You'll always have some resistance there, because there are no perfect inductors, but for the sake of argument, let's remove the resistor for a moment.

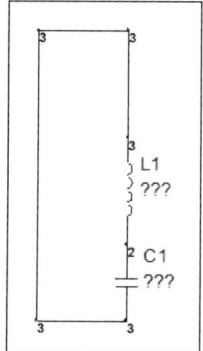

So…is L1 in series or in parallel with C1? Don't think too long; it's a simple question.

Stumped?

Well, it *looks* like a series circuit, you say. Something about it just looks "series-y."

Okay…let's help you out a bit. Why don't we just redraw this a tad.

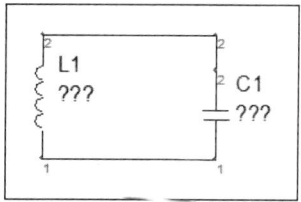

As you can see, nothing has changed, electrically. It's just mechanically drawn to look more "parallel-y." The truth of the matter is, there is absolutely no difference between a series and parallel resonant circuit. One thing we do know about this. We will have *maximum* circulating current at resonance. In fact, unless we do have some resistance in there, our circulating current will be infinite.

But, how do we get current to circulate in the first place? We need an ac voltage source. There are actually two ways we can connect up our voltage source (V1 in the drawing). We can do it like this.

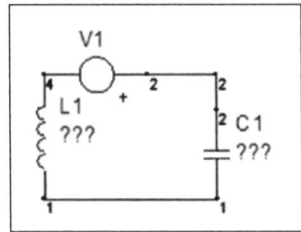

Or we can do it like this.

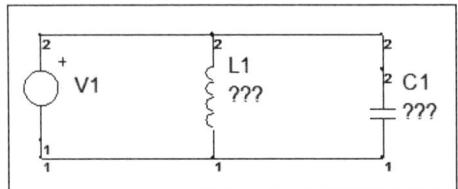

Now — and *only* now — do we have a difference between "series resonance" and "parallel resonance." It's defined by how we connect up our power source!

Let's look at our first case. It is clearly a series resonant circuit now. There is only one path around the circuit, so the currents in all three nodes have to be the same. We will find that our voltage source has maximum current at the resonant frequency.

Now, let's look at our parallel hookup. At resonance, the circulating current between L1 and C1 is maximum, as in the first case. But how about the current passing through the generator (V1)? At resonance, this current will be *zero*!

How can this be? Well, you either believe Kirchhoff's Current Law or you don't. Let's look at Node 2, where we actually have three connections. If we have infinite current going *up* through L1, and infinite current going *down* through C1, how much current is allowed into that node from another source (our generator). *None*! So we find that when we have the greatest *circulating* current, we have the minimum *driving* current! For those of you familiar with vacuum tube amplifiers, this explains why you have the maximum output power when the plate current dips. The plate of the tube is basically a voltage source, while the Pi-network circulating current is what's actually coupled to the antenna.

Now, as it turns out, both *SPICE* programming and real circuits have difficulties dealing with infinite values. So let's put our original resistor back in, just to keep our numbers sane.

And…for the remainder of this discussion, we can ignore our obvious series resonant circuit, since it is really irrelevant to the matter of loaded and unloaded Q….which is what the entire discussion up to this point is setting us up for!

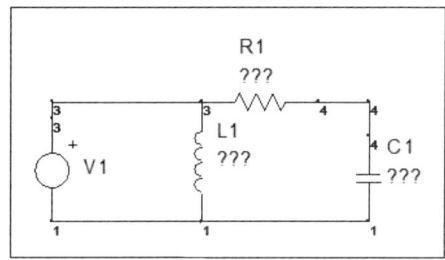

In the figure directly above, we see a typical *unloaded* tank circuit, such as you might find in the final stage of a tube amplifier. The R we see in this circuit is the value that determines the *unloaded* Q of the circuit. In fact, this R is the limiting factor for the overall efficiency of the tank circuit. At this point in time, we can ignore the contribution of the voltage source to the total current. But, unlike in the previous example, there *is* a finite amount of current being drawn from the source. As the value of R approaches zero, the current drawn from V1 approaches zero, as well. Oh….and the formula for Series Q fully applies. Q simply = X/R

Now, as interesting as this circuit is, it really doesn't do much for us. There's no load. We're just circulating a bunch of "wattless" power around the L1/C1 loop, minus some very small (we hope), power loss in R1.

To actually do something with this circuit, we need to add a load, which we will call R2. R2 can be an antenna, a dummy load, or perhaps the input to another power amplifier stage.

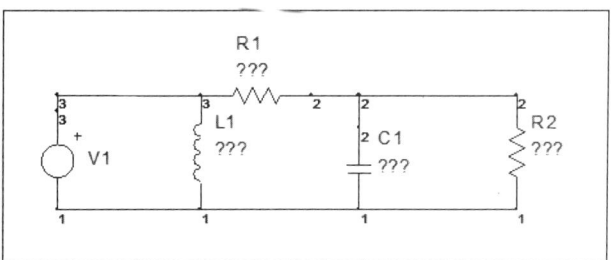

Now, if we, for a moment, disregard L1 and C1, can you see that R1 and R2 form a voltage divider? The *efficiency* of the network is directly related to the ratio of R2 to R1. If R1 were zero, our entire power would be dissipated in R2 (which would be our ultimate goal). In this case our system is 100% efficient. Wouldn't that be lovely?

Now, mentally reinsert L1 and C2. We can also see that R2 is going to have some effect on the overall Q of the circuit. It's going to dissipate power. And it doesn't really matter *how* we dissipate power — either through R1 or R2, the results on Q are going to be much the same. We are going to not recover the entire circulating current.

Role Reversal

If you're like most hams, you've probably been a bit baffled by the assertion that for a "parallel" resonant circuit, Q is now defined by R/X. That's the reciprocal of what we consider normal. To address the confusion, we need to understand *which* R is being referred to! In a *heavily loaded* circuit (one in which R2 is small relative to the reactance of L1 or C2), *R2* is the "R" of the Q equation. Not so, for the "unloaded" version, where R1 dominates. It should be fairly evident that if R2 is an extremely high value, it will have minimal effect on the circuit, under which case the Q will be high. Thus Q = R2/X.

As Easy as Pi

There's nothing particularly mystical about the Pi network; all the rules about loaded and unloaded Q still apply. But it's an interesting example, worth a bit of investigation.

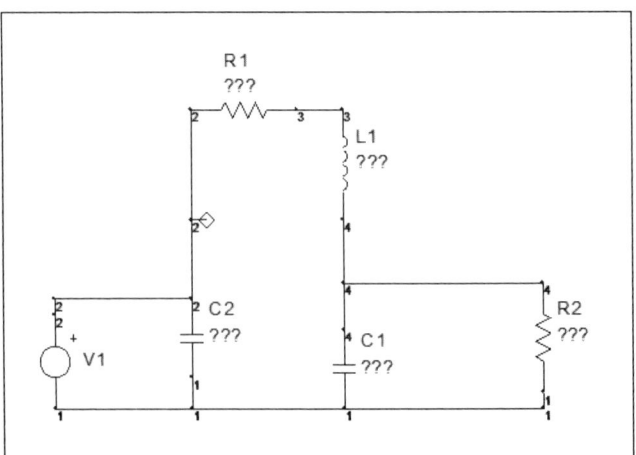

We hope you see by now how "mechanically" redrawing an electrical circuit can clear up a lot of confusion. In the same vein, the diagram above is identical to the normal Pi network, just with a different physical orientation of L1, for a reason that will become very clear. R1 again, is the internal loss resistance, primarily responsible for the *unloaded* Q, while R2 is the load, responsible for the *loaded* Q value.

C2 is the normal "tuning" capacitor, which is actually in *series* with the "loading" capacitor C1. The total C can never be more than the value of C2, and so C2, in conjunction with L1, primarily determines the resonant frequency of the entire "tank." C1 is usually many times as large as C2. We also see that L1 and C1 form a voltage divider. The voltage appearing across R2, the load resistor, is proportional to the ratio of C1 reactance to L1 reactance. We therefore have control of the voltage (and power) dissipated in R2 by means of adjusting the L1/C1 ratio. Since the resonant frequency is mainly determined by C2, C1 can be changed a bit without significantly "detuning" the circuit.

Changing the value of C1 not only changes the output power, but it changes the Q of the circuit. The *loaded* Q, to be specific. When C1 is very large, its reactance is small relative to R2, which means high loaded Q.

Now, a few words about R1. We don't have much control over R1; it is built into the losses, primarily of L1. R2 isn't normally adjustable either, it is the load impedance of the antenna that we're dealt. However, L1 and C2 also serve as an L network, which can *transform* the effective resistance of R2...just as any L network can. The way we maintain the *efficiency* of the whole, is to assure that *effective value* or *transformed* value of R2 is a much higher value than R1. As you can see, there is a bit of a balancing act between L and C values for best efficiency.

In Circulation

We have clearly shown (we hope) that KCL allows us to have very different circulating and input current values, giving us the peculiar properties and benefits of the so-called parallel resonant circuit. We have violated no electrical laws in any of this discussion.

Appendix C

Original article by Dr. Mike Trimpi

Broadband Loop Antenna System

MLT Jan 27, 2014

Broadband Loop Antenna System

The following assumptions are made:

1. The circumference times the number of turns is small compared compared to one wavelength at the highest frequency of interest.
2. The DC resistance of the loop $<< \Omega L$ at the lowest frequency of interest.
3. Any parasitic effects are ignored.

The basic circuit is:

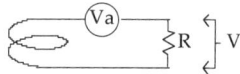

Response to a received signal:

$Va = E\,A\,n\,\omega/c$ Va = Voltage induced in loop by the received signal
$V = Va\,R/((\omega\,L)^2 + R^2)^{1/2}$ E = Strength of received signal (V/M)
A = Area of loop (M^2)
n = Number of turns
$\omega = 2\pi f$
c = Speed of light (3×10^8 M/Sec)
L = Loop inductance (H)

At low frequencies where ωL is small compared to R the loop looks like a voltage source so the voltage across R is proportional to the received frequency (ω). At frequencies where ωL is large compared to R, Va and ωL are both increasing with frequency so the antenna looks like a constant current source. Since R is constant, the response in this frequency range is flat. Thus, V rises with frequency at a rate of 6 dB/octave until ωL approaches R when the response flattens and V remains constant. At the frequency where $\omega L = R$ the response is down by 3 dB.

In most applications the loop impedance is very low compared to the optimum source impedance of the amplifier so a transformer is placed between the loop and the amplifier. Zp, the impedance looking into the transformer primary replaces R so the primary circuit is now ωL in series with Zin/N^2 where N is the secondary/primary turns ratio.

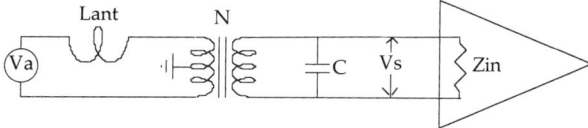

Vs, the voltage at the transformer secondary, is:

$Vs = Ip\,Zin/N$ N = Secondary to primary turns ratio
$Ip = Va/((\omega\,L)^2 + Zp^2)^{1/2}$ Zp = Impedance looking into the primary
$Zp = Zin/N^2$ Zin = Input impedance of amplifier.
$Vs = Vp\,N$

MLT Jan 27, 2014

A capacitor can be connected across the transformer secondary to provide high frequency attenuation. Strictly speaking, this capacitor, the antenna inductance, and Zin form an RLC resonant circuit in which the L and C determine the resonant frequency and Zin determines the Q. However, in this application the Q is very low, typically less than .3, so the high pass function of ωL vs Zin/N^2 and the lowpass function of $1/\omega C$ vs Zin can be calculated independently. The transformer secondary is now loaded by the parallel impedance of Zin and $1/\omega C$. At high frequencies $N^2 \omega L$, the impedance looking back into the transformer secondary, is now much higher than Zin so the upper -3 dB frequency occurs where $1/\omega C = Zin$.

At mid-band frequencies where $\omega L N^2 < Zin < 1/\omega C$, Vs can be approximated by:

Vs = E A n / L N c

Calibration:

The simplest method of applying a calibration signal is by the constant current method. If Rcal is very large compared to the primary circuit impedance, the current through Rcal is independent of the calibration frequency and is, therefore, constant.

The voltage across Rcal then equals Vcal and the calibration current equals:

Ical = Vcal / Rcal
Vp = .5 Ical (Zp $\parallel \omega L$)
Zp = (Zin $\parallel 1/\omega C$) / N^2
Vs = Vp N

The factor of .5 accounts for the transformer primary center-tap being grounded to reduce sensitivity to electric fields. The impedance looking into 1/2 the primary is 1/4 of the whole so the primary voltage is 1/4 but the turns ratio is now 1:10 so Vs is 1/2 of a non-center tap circuit. Midband approximation: The calculations can be simplified when a frequency is chosen that is between the lower and upper 3 dB frequencies. In this case $\omega L \gg Zin/N$ and $1/\omega C \gg Zin$ so:

Vs = Vcal Zin / 2 N Rcal

C.2 Appendix C

MLT Jan 27, 2014

The noise generated by Rcal is insignificant because its source impedance, Rcal itself, is much greater than the load impedance, Zp $\|$ ω L. The noise generated by a resister of 39 K is about 25 nV. For N = 5 and Zin of 1500 Ω the noise at the amplifier input caused by Rcal equals .2 nV which is small compared to the amplifier noise.

Calibration signals:
  Sine wave: Accurate amplitude calibration but at only one frequency.
  Swept sine wave: Choosing a dF/dt may be a problem depending on data analysis.
  White noise: Amplitude measurement requires a spectrum analyzer.
  Pseudorandom noise generator: Generates a comb spectrum. Amplitude measurement requires a spectrum analyzer but the amplitude can be checked for consistancy with an oscilloscope.

Minimum Detectable Signal (MDS):

If the MDS is defined as a signal equal to the amplifier noise, Vn, then:

$$\text{MDS (V/M)} \equiv Vn\ N\ L\ c / A\ n\ Zin$$

The antenna/preamp system often used with the Dartmouth LF-HF receiver has the following parameters:

A = 10 M^2
n = 1
Lant = 18 uH
Rcal = 39 K   Value chosen so E/Vcal = .01
N = 5
C = 18 pF   Chosen to give increasing attenuation above about 5 MHz
Zin = 1800 Ω   Input impedance of amplifier
Rfb = 360 K Ω   Sets Zin; =360 K/200 = 1800 Ω

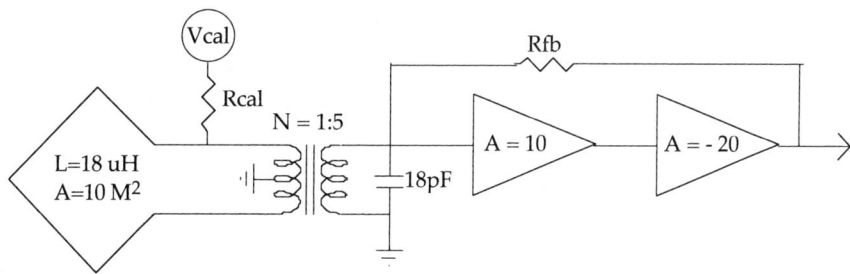

The preamplifier consists of two op - amps. The first is configured in the non-inverting mode, the second in the inverting mode. Zin, the input impedance of the amplifier = Rfb/total gain. This

MLT Jan 27, 2014

configuration generates less noise than a resistor across the transformer secondary.
Assuming perfect components, this antenna/preamp system will have the following characteristics:
 Lower -3dB frequency is that at which $Zin/N^2 = \omega L = 160$ KHz
 Upper -3dB frequency is that at which $1/\omega C = Zin = 5.0$ MHz

Minimum Detectable Signal: $(Va \times N = Vn) = 15$ nV/M/√Hz
Vout due to preamp noise $= .2$ μV/√Hz
Vout (V) $= 13.3$ E (V/M)
Vcal (V) $\equiv 9.6$ mV/(M/√Hz)
Vout $= .13$ Vcal

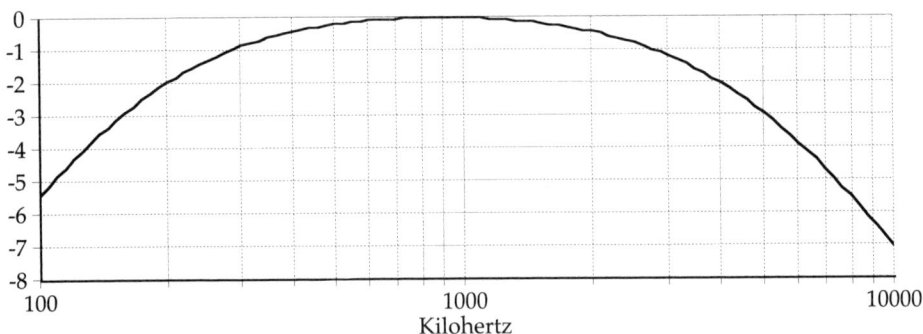

Response (dB) vs Frequency (KHz)

Choice of amplifiers: For a given antenna system the sensitivity will depend on the first stage noise plus any possible noise generated by the antenna and transformer resistance. If possible, choose an op-amp having an optimum source impedance (Op Z) equal to Rfb/total gain. The true source impedance is Rfb/gain in parallel with ΩLN^2. However, above the lower -3dB frequency Rfb/gain dominates. If possible, a transformer is chosen that gives the desired lower -3dB frequency for the optimum impedance. Often, for a given antenna, it is impossible to find the ideal transformer. Little is lost by changing Rfb to give the desired frequency response.

The following is a short list showing typical noise values of some low noise amplifiers.
Noise power = Vnoise x Inoise; Optimum Z = Vnoise/Inoise.

Amp	Vnoise (nV)	Inoise (pA)	Noise pwr (W^{-21})	Op Z (KΩ)	Gain x BW (MHz)
LT1028	1.1	0.9	1.0	1.3	50
INA217	1.3	0.8	1.0	1.6	see data sheet
LT1115	0.9	1.2	1.1	0.8	40
OP37	3.0	0.4	1.2	7.5	60
HA5137	3.0	0.4	1.2	7.5	63
HA5147	3.2	0.4	1.3	8.0	120
EL5135	1.5	0.9	1.3	1.7	650

C.4 Appendix C

Appendix D: Low Noise Amplifiers

Device	Vnoise (nV)	Inoise (pA)	Op Z (kΩ)	Noise Pwr (W^{-21})	GBP	Package	Mfgr.	Notes
LMC6001	22	0.00012	183333	0.003	1.3M	DIP	Nat. Semi.	Zin = 1 TΩ P-P supply = 16 V
LTC6240	7	0.0006	11667	0.004	18M	SOT23, SO	Lin.Tech.	Single & dual
OPA827	4	0.0022	1818	0.009	22M	SO-8	Tex. Inst.	
2SK170	0.95	0.01	95	0.009		TO-92	Toshiba	discrete FET
LSK170	0.95	0.01	95	0.009		TO-92	Lin. Sys.	discrete FET
OPA376	7.5	0.002	3750	0.015	55M	SOIC, SOT23, SC70	Tex. Inst.	dual
AD743	2.9	0.007	414	0.020	4.5M	14 pin DIP	Ana. Dev.	
AD745	2.9	0.007	414	0.020	20M	14 pin DIP	Ana. Dev.	
LT1792	4.2	0.01	420	0.042	5.6M	DIP, SO	Lin. Tech.	
LM6211	5.5	0.01	550	0.055	20M	SOT23	Nat. Semi.	
LT1024	14	0.006	2333	0.08	1M	14 pin DIP		Dual amp
INA163	1	0.8	1.2	0.80	800K	14 pin SM	Tex. Inst.	Instrument. amp
LT1124	2.7	0.3	9.0	0.81	12M		Lin. Tech.	
LT6233	1.9	0.43	4.4	0.82	60M		Lin. Tech.	
AD620	9.0	0.1	90	0.90	10M			Vn gain dependent
LT1028	1.1	0.85	1.3	0.94	50M	DIP, SOIC	Lin. Tech.	
LT1007	2.5	0.4	6.2	1.00	60M	DIP	Lin. Tech.	
LT6237	2.5	0.4	6.2	1.00	60M		Lin. Tech.	
INA217	1.3	0.8	1.6	1.04		SOIC, DIP	Burr-Brown	See data sheet for GBP
LT1115	0.9	1.2	0.8	1.08	40M	DIP	Lin. Tech	
LT1001	10	0.11	91	1.10	800K	DIP	Lin. Tech	
LT6230	1.1	1	1.1	1.10	215M	SOT-23	Lin. Tech	
OP77	9.6	0.12	80	1.15	600K	DIP		
HA5137	3	0.4	7.5	1.20	63M	8 pin CerDIP	Intersil	
OP27	3	0.4	7.5	1.20	8M	DIP		
OP37	3	0.4	7.5	1.20	60M	DIP		
OPA228	3	0.4	7.5	1.20	33M			
HA5147	3.2	0.4	8.0	1.28	120M		Intersil	
EL5135	1.5	0.9	1.7	1.35	650M	SOT23	Intersil	
NE5533	3.5	0.4	8.8	1.40	10M			
LT6233-10	1.9	0.78	2.4	1.48	375M	SOT23	Lin. Tech.	
OPA129	15	0.1	150	1.5	1.3M	DIP	Burr-Brown	Zin 10 TW
AD8429	1	1.5	0.7	1.50	1.2 MHz @ A = 100		Ana. Dev.	V noise and BW gain dependent

AD8099	0.95	1.6	0.6	1.52	500M	SOIC, CSP		Ana. Dev.	
LT6202	1.9	0.8	2.4	1.52	100M	SOT23, SO		Lin. Tech.	Check datasheet for noise values
AD8597	1.1	1.5	0.7	1.65	10M	8SOIC		Ana. Dev.	
AD8432	0.85	2	0.4	1.70	200M	24 lead		Ana. Dev.	Prog. gain/active Zin
SSM2134	3.5	0.5	7.0	1.75	10M	DIP		Ana. Dev.	
LHM6629	0.69	2.6	0.3	1.79	200M				
MAX410	1.5	1.2	1.2	1.80	28M	DIP, SO		Maxim	
AD797	0.9	2.0	0.5	1.80				Ana. Dev.	GBP: 110MHz @ A = 1000; see DS
LMH6626	1	1.8	0.6	1.80	1.5G	SOT23			
LM6629	0.7	2.6	0.3	1.82		SOT23		Nat. Semi.	fast, see datasheet / 15 mA
OPA211	1.1	1.7	0.6	1.87	50M			Tex. Inst.	
MAX4106	0.75	2.5	0.3	1.88	350M			Maxim	
EL1516	1.3	1.5	0.9	1.95	350M		8SIOC	Intersil	
EL6131	1.8	1.1	1.6	1.98	300M			Intersil	
ISL28134	10	0.2	50	2.00	3.5M				Chopper stabilized 1/Fo = 0.1 Hz
AD117	118	3	39	354	.6M	DIP		Ana. Dev.	Very low offset - very noisy

D.2 Appendix D

Appendix E: Online Resources

Here are some handy on-line calculators and other references. There was a time when every well-stocked ham shack had a wide variety of charts and nomographs. Nowadays there are so many wonderful calculators online that save a lot of dead trees. (I do however, still have my students work out Smith Chart problems with a compass and straight edge.) In any case, here are some links to some great on-line calculators, particularly useful for receiving antenna design work.

electronbunker.ca/eb/InductanceCalcML.html

I would hate to have to do calculations for multilayer inductors longhand. This calculator is great for figuring out close-wound multilayer receiving loops and such.

home.earthlink.net/~jimlux/hv/wheeler.htm

Not exactly a calculator, but a classic review of the Wheeler inductor. For a given length of wire a particular ratio of length to diameter of a coil will give you the maximum inductance (which usually results in the highest Q). Wheeler derived this formula decades ago, and it has only been slightly improved upon. The Wheeler coil is approximately square (just slightly shorter than its diameter).

www.deephaven.co.uk/lc.html

A really basic resonant frequency calculator for the very lazy.

www.easycalculation.com/physics/electromagnetism/wire-resistivity-calculator.php

A wire resistance calculator. If you plan on exploring our new "top band" at 2200 meters, wire resistance of tuned loops becomes a factor. The Q of your tuned loop is going to be largely determined by the wire resistance.

makearadio.com/tech/files/Ferrite_Rod_Inductance.pdf

This is the best paper I've found on calculating the inductance of a ferrite loopstick. Increasingly more relevant with our new lower bands.

mustcalculate.com

A large collection of online electronics calculators.

Index

The letters "ff" after a page number indicate coverage of the indexed topic on succeeding pages.

2200 meter band: 16.1, 24.1ff
4nec2: 4.4, 7.4, 11.7
630 meter band: 16.1, 24.1ff

A

Active antenna: 4.2, 6.1ff
   Power supply: 19.1ff
   Voltage probe: 5.7
Adcock antenna: 9.10, 22.26ff
Analog Devices AD8067 op-amp: ...6.3, 10.8ff, 22.4
Analog Devices AD8307 log-amp: 10.8ff, 22.6
Angle of arrival: 21.3
Antenna
   Active: 4.2
   Adcock: 22.26ff
   Aperiodic: 5.4, 11.1ff
   As a signal generator: 5.1ff
   Beverage: 13.1ff
   Beverage on Ground (BOG): 13.3
   Circular polarization (CPOL) diversity
      array: 22.32ff
   eXOgon: 15.1ff, 22.17ff
   Full wave loop: 8.2
   Helical frame loop: 22.10ff
   Loopstick: 12.7
   Low-band receiving loop: 22.6ff
   Multi-turn loop: 12.6
   Near Vertical Incident Skywave (NVIS): ...22.21ff
   Random wire: 17.1ff
   Resonant: 5.3
   Shielded ferrite loopstick for 630 meters: .22.22ff
   Small: 4.1ff
   Small loop: 8.1ff
   Small tuned loop (STL): 4.1
   Trimpi Loop: 22.15ff
   Very short: 4.4ff
   Wave: 13.2
Antenna analyzer: 5.5
Antenna effect: 9.10
Antenna tuner (adjusting phasing with): 18.8

Aperiodic antenna: 5.4
Aperiodic loop: 7.3, 11.1ff
ARRL 600 Meter Experimental Group: ... 16.1, 24.1
Attenuator: 10.7

B

Balun: 17.12, 19.3
Beamforming network: 18.1ff
Beverage antenna: 13.1ff
   At 630 and 2200 meters: 24.3
   Short: 13.5
Beverage on Ground (BOG): 13.3
Bi-refringent: 1.1
Bias T: 19.2
   Bipolar power supply: 19.4
Butler Matrix: 18.3, 22.21ff

C

Capture area: 1.3
Cardioid pattern: 9.5
Circular polarization (CPOL): 14.4, 15.2ff, 20.6
   Diversity array: 22.32ff
Components: 23.10
Conjugate match: 5.4
Construction techniques: 23.1ff
Cosmic noise: 3.2

D

dBm: 2.6
Decibel: 2.1ff
   Power ratio: 2.2
   Referenced to 1 mW (dBm): 2.6
   Voltage-derived: 2.3ff
Digisonde: 21.3, 21.8
Diversity reception: 20.1ff

E

Efficiency: 1.2
End-fed half-wave (EFHW) antenna: 17.12
EWE: 7.3, 21.4
eXOgon antenna: 15.1ff, 22.17ff
   Construction details: 15.6ff
Extended double Zepp (EDZ): 1.5

F

Ferrite: 12.7, 16.4, 22.22ff
Field effect transistor (FET) amplifier: 6.1ff
Field strength measurements: 10.1ff
Field strength meter: 10.3, 10.7ff, 22.2ff
Folded dipole: 7.2
Frequency diversity: 20.4
FT8: 10.5, 16.2

G

Ground conductivity: 14.2, 16.6, 17.8
Ground wave propagation: 8.3

H

HAARP: 21.3, 21.5
Harmonic antenna: 11.1
Helical frame loop: 22.10ff
HIPAS Observatory: 15.1ff, 18.1, 21.5, 21.6

I

Insulators: 23.5
Interference: 20.3
Intermodulation distortion (IMD): 12.3
Ionogram: 21.8
Ionosonde: 21.8

J

JT65, JT9: 10.5, 16.2, 24.2

K

K9AY Loop: 7.3, 11.5ff, 21.4
    Model: 11.7ff
Kirchoff's Current Law (KCL): B.1ff

L

Litz wire: 16.5
Loaded Q: B.1ff
Logarithmic amplifier (log-amp): 10.11ff
    Analog Devices AD8307: 10.8ff, 22.6
Loopstick antenna: 12.7
    At 2200 meters: 24.4ff
    For 630 meters: 22.22ff, 24.4ff
    Null: 9.9
Loose coupler: 8.2
Low frequency (LF): 16.1ff
Low noise amplifier devices: D.1ff
Low-band receiving loop: 22.6ff

M

Magnet wire: 23.1
Mini-Circuits JSPQ-65W+ hybrid: 11.5, 15.5

Model
    K9AY Loop: 11.7
    Random wire antenna: 17.4ff
    Very short antenna: 4.4ff
Modeling: 7.4

N

Near Vertical Incident Skywave (NVIS): 21.1ff
    Antenna project: 22.21ff
    At 630 and 2200 meters: 24.6
Neper: 2.2
Noise: 1.5, 3.1, 12.5
    Contribution from preamplifier: 3.3
    Polarization of: 8.4
    Thermal: 3.1, 7.5
Norton-Podell amplifier: 16.9
Null: 4.2, 9.1ff

O

Op-amp.. 6.3
    Analog Devices AD8067: 6.3, 10.8ff, 22.4

P

Pennant: 7.3, 21.4, 22.21ff
Phased array: 18.1ff
Power: 2.2
Power factor: 5.7
Power transfer: 5.2
Preamplifier: 3.1ff, 6.1ff
    AD8067 op-amp: 6.3
    Effect on system noise: 3.3
    Low-noise devices D.1ff
    Protection: 19.5
    Undesirable effects: 3.3
Preselector: 3.1, 3.3
Printed circuit board (PCB): 23.6
Project boxes: 23.2
Projects: 22.1ff
Propagation, non-reciprocal: 14.1ff
PropLab Pro software: 14.3
PVC plumbing: 23.6

Q

QRSS (Very Slow CW): 16.8
Quality factor (Q): 12.1ff, A.1ff, B.1ff

R

Radiation resistance:........................ 5.2, 7.2, 23.12
Radio direction finding (RDF): 4.2, 8.5, 9.1ff
    Low band: 9.8
Random wire antenna: 17.1ff
    Model: 17.4ff
Receive antenna input: 19.6
Received signal levels: 1.3

Receiver
    Input impedance measurement: 5.4ff
    Overload: 3.3
    Sensitivity test: 3.2
Reciprocity: 1.1
Resistive loading: 7.1ff
Resonance: 5.7, 11.4, 12.4, B.1ff
Resources: E.1ff
Reverse Beacon Network (RBN): 10.5
Riometer: 3.2
Rope: 23.4

S

S-9 signal level: 2.6
S-meter: 1.3, 3.3, 10.2
    For direct conversion receivers: 22.30ff
Select-A-Tenna: 22.10
Selective fading: 20.2
Sense antenna: 8.5, 9.4, 22.28
Shielded ferrite loopstick for 630 meters: 22.22ff
SI (International System) units: 2.1
Signal levels, received: 1.3
Signal-to-noise: 1.5
Small antenna: 4.1ff
    Advantages: 4.2
Small loop antenna: 8.1ff
    Noise reduction: 8.4
    Null: 9.6ff
    Polarization of: 8.4
Small tuned loop (STL): 4.1
Soldering equipment: 23.6
Storage cabinets: 23.10
SWR bandwidth: 12.2
Synthetic aperture antenna: 18.4, 18.6

T

Test equipment: 23.7
Thermal noise... 3.1
    At VHF/UHF: 3.3
Tickler coil: 16.6
Time diversity: 20.4
Tools: 23.7
Total Copper Content (TCC): 1.4
Transmission line: 23.4
Trimpi Loop: 11.3, 16.9, 22.15ff, C.1ff
Tubing and rod: 23.6

U

Unloaded Q: B.1ff

V

Vector network analyzer (VNA): 5.4

W

Waller Flag: 7.3
Wave antenna: 13.2
WSJT: 10.5
WSPR: 10.5
Wullenweber antenna: 18.4

X

X and O waves: 14.4, 15.3, 20.6

F
Ferrite: 12.7, 16.4, 22.22ff
Field effect transistor (FET) amplifier: 6.1ff
Field strength measurements: 10.1ff
Field strength meter: 10.3, 10.7ff, 22.2ff
Folded dipole: 7.2
Frequency diversity: 20.4
FT8: 10.5, 16.2

G
Ground conductivity: 14.2, 16.6, 17.8
Ground wave propagation: 8.3

H
HAARP: 21.3, 21.5
Harmonic antenna: 11.1
Helical frame loop: 22.10ff
HIPAS Observatory: 15.1ff, 18.1, 21.5, 21.6

I
Insulators: 23.5
Interference: 20.3
Intermodulation distortion (IMD): 12.3
Ionogram: 21.8
Ionosonde: 21.8

J
JT65, JT9: 10.5, 16.2, 24.2

K
K9AY Loop: 7.3, 11.5ff, 21.4
  Model: 11.7ff
Kirchoff's Current Law (KCL): B.1ff

L
Litz wire: 16.5
Loaded Q: B.1ff
Logarithmic amplifier (log-amp): 10.11ff
  Analog Devices AD8307: 10.8ff, 22.6
Loopstick antenna: 12.7
  At 2200 meters: 24.4ff
  For 630 meters: 22.22ff, 24.4ff
  Null: 9.9
Loose coupler: 8.2
Low frequency (LF): 16.1ff
Low noise amplifier devices: D.1ff
Low-band receiving loop: 22.6ff

M
Magnet wire: 23.1
Mini-Circuits JSPQ-65W+ hybrid: 11.5, 15.5

Model
  K9AY Loop: 11.7
  Random wire antenna: 17.4ff
  Very short antenna: 4.4ff
Modeling: 7.4

N
Near Vertical Incident Skywave (NVIS): 21.1ff
  Antenna project: 22.21ff
  At 630 and 2200 meters: 24.6
Neper: 2.2
Noise: 1.5, 3.1, 12.5
  Contribution from preamplifier: 3.3
  Polarization of: 8.4
  Thermal: 3.1, 7.5
Norton-Podell amplifier: 16.9
Null: 4.2, 9.1ff

O
Op-amp... 6.3
  Analog Devices AD8067: 6.3, 10.8ff, 22.4

P
Pennant: 7.3, 21.4, 22.21ff
Phased array: 18.1ff
Power: 2.2
Power factor: 5.7
Power transfer: 5.2
Preamplifier: 3.1ff, 6.1ff
  AD8067 op-amp: 6.3
  Effect on system noise: 3.3
  Low-noise devices D.1ff
  Protection: 19.5
  Undesirable effects: 3.3
Preselector: 3.1, 3.3
Printed circuit board (PCB): 23.6
Project boxes: 23.2
Projects: 22.1ff
Propagation, non-reciprocal: 14.1ff
PropLab Pro software: 14.3
PVC plumbing: 23.6

Q
QRSS (Very Slow CW): 16.8
Quality factor (Q): 12.1ff, A.1ff, B.1ff

R
Radiation resistance:.................... 5.2, 7.2, 23.12
Radio direction finding (RDF): 4.2, 8.5, 9.1ff
  Low band: 9.8
Random wire antenna: 17.1ff
  Model: 17.4ff
Receive antenna input: 19.6
Received signal levels: 1.3

Receiver
    Input impedance measurement: 5.4ff
    Overload: 3.3
    Sensitivity test: 3.2
Reciprocity: 1.1
Resistive loading: 7.1ff
Resonance: 5.7, 11.4, 12.4, B.1ff
Resources: E.1ff
Reverse Beacon Network (RBN): 10.5
Riometer: 3.2
Rope: 23.4

S

S-9 signal level: 2.6
S-meter: 1.3, 3.3, 10.2
    For direct conversion receivers: 22.30ff
Select-A-Tenna: 22.10
Selective fading: 20.2
Sense antenna: 8.5, 9.4, 22.28
Shielded ferrite loopstick for 630 meters: 22.22ff
SI (International System) units: 2.1
Signal levels, received: 1.3
Signal-to-noise: 1.5
Small antenna: 4.1ff
    Advantages: 4.2
Small loop antenna: 8.1ff
    Noise reduction: 8.4
    Null: 9.6ff
    Polarization of: 8.4
Small tuned loop (STL): 4.1
Soldering equipment: 23.6
Storage cabinets: 23.10
SWR bandwidth: 12.2
Synthetic aperture antenna: 18.4, 18.6

T

Test equipment: 23.7
Thermal noise..3.1
    At VHF/UHF: 3.3
Tickler coil: 16.6
Time diversity: 20.4
Tools: 23.7
Total Copper Content (TCC): 1.4
Transmission line: 23.4
Trimpi Loop: 11.3, 16.9, 22.15ff, C.1ff
Tubing and rod: 23.6

U

Unloaded Q: B.1ff

V

Vector network analyzer (VNA): 5.4

W

Waller Flag: 7.3
Wave antenna: 13.2
WSJT: 10.5
WSPR: 10.5
Wullenweber antenna: 18.4

X

X and O waves: 14.4, 15.3, 20.6